T0310818

Towards Deep Understanding of

Elementary School Mathematics

A Brief Companion for Teacher Educators and Others

Other World Scientific Titles by the Author

Exploring Mathematics with Integrated Spreadsheets in Teacher Education
ISBN: 978-981-4678-22-3
ISBN: 978-981-4689-90-8 (pbk)

Diversifying Mathematics Teaching: Advanced Educational Content and Methods for Prospective Elementary Teachers
ISBN: 978-981-3206-87-8
ISBN: 978-981-3208-90-2 (pbk)

Integrating Computers and Problem Posing in Mathematics Teacher Education
ISBN: 978-981-3273-91-7

Towards Deep Understanding of

Elementary School Mathematics

A Brief Companion for Teacher Educators and Others

Sergei Abramovich
State University of New York at Potsdam, USA

NEW JERSEY · LONDON · SINGAPORE · BEIJING · SHANGHAI · HONG KONG · TAIPEI · CHENNAI · TOKYO

Published by

World Scientific Publishing Co. Pte. Ltd.
5 Toh Tuck Link, Singapore 596224
USA office: 27 Warren Street, Suite 401-402, Hackensack, NJ 07601
UK office: 57 Shelton Street, Covent Garden, London WC2H 9HE

Library of Congress Cataloging-in-Publication Data
Names: Abramovich, Sergei, author.
Title: Towards deep understanding of elementary school mathematics :
a brief companion for teacher educators and others /
Sergei Abramovich, State University of New York at Potsdam, USA.
Description: New Jersey : World Scientific, [2022] | Includes bibliographical references and index.
Identifiers: LCCN 2022013199 | ISBN 9789811256998 (hardcover) |
ISBN 9789811257001 (ebook for instituations) | ISBN 9789811257018 (ebook for individuals)
Subjects: LCSH: Mathematics--Study and teaching (Elementary)
Classification: LCC QA135.6 .A23 2022 | DDC 372.7--dc23/eng20220617
LC record available at https://lccn.loc.gov/2022013199

British Library Cataloguing-in-Publication Data
A catalogue record for this book is available from the British Library.

For any available supplementary material, please visit
https://www.worldscientific.com/worldscibooks/10.1142/12865#t=suppl

Typeset by Stallion Press
Email: enquiries@stallionpress.com

Printed in Singapore

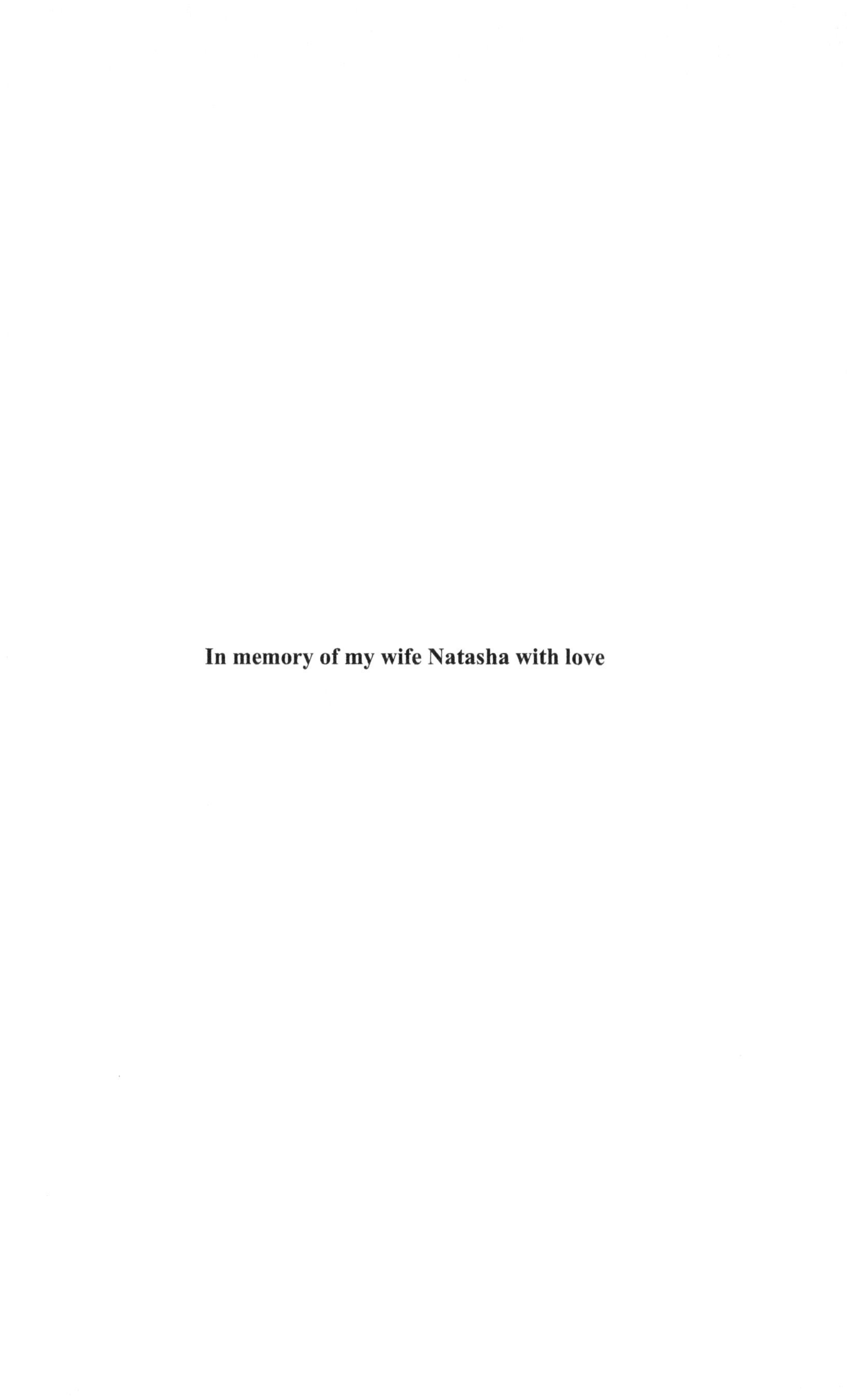

In memory of my wife Natasha with love

Preface

The textbook is written to serve as a brief companion for mathematical educators of K-6 teacher candidates who learn elementary mathematics methods and content both at the undergraduate and graduate levels. The applicability of mathematical content of the textbook to a teacher preparation program stems from the content being supported by a number of notable modern-day educational documents used in the United States such as Common Core State Standards (http://www.corestandards.org), NCTM Principles and Standards for School Mathematics (https://www.nctm.org), AMTE Standards for Preparing Teachers of Mathematics (https://amte.net), and the Conference Board of the Mathematical Sciences (https://www.cbmsweb.org) recommendation for the preparation of mathematics teachers. To make the textbook applicable to teacher preparation programs outside the United States, the above list has been extended to include such documents as National Curriculum (England), National Curriculum in Mathematics (Australia), Primary Mathematics Teaching and Learning Syllabus (Singapore), The Ontario Curriculum (Canada), Elementary School Teaching Guide (Japan), National Mathematics Curriculum (Korea), Primary Mathematics Standards (Chile), Mathematics Teaching and Learning Framework (South Africa), and other documents with which the author is familiar from previous writings. In that way, the textbook can serve international readership.

The basic didactic ideas of the textbook include the development of the concepts of positive integers (whole numbers) and base-ten system; the emergence of fractions in the context of simple real-life activities requiring the extension of integer arithmetic; decimals, percent and ratio; geoboard geometry and activities in support of the van Hiele model of geometric thinking; combinatorics; probability and data analysis. The textbook emphasizes the importance of deep conceptual understanding of elementary school mathematics by prospective elementary teachers. The need for deep understanding is at least twofold. First, the modern-day students have been actively seeking answers to conceptual questions about procedural knowledge they gain through the study of school mathematics. Second, as mathematics educators worldwide agree, teachers of mathematics must know content they teach both from an advanced perspective and beyond the level of their current teaching responsibilities.

The content of the textbook fits in a three credit hours course. It has been informed by the author's participation in field-based research grounded in supervising teacher candidates' work with young children in regular classrooms that often extended to various after school learning initiatives. Consequently, many articles published by the author over the years as reflection on those initiatives inform the core content of the textbook.

The textbook includes historical aspects of elementary school mathematics. For example, two-sided counters stem from the binary system with its genesis in the 1st millennium B.C. China of which Leibnitz was one of the first notable proponents; the genesis of the base-ten arithmetic is in the Egyptian mathematics of the 4th millennium B.C., enriched by the positional notation with the advent of Hindu-Arabic numerals in the 12th century Europe. An emphasis of the Common Core State Standards on visualization through the use of the so-called tape diagrams goes back to Aristotle's recognition of the power of visual thinking. Other historical connections and cultural origins of elementary mathematics are highlighted throughout the textbook including the names of Confucius, Pythagoras, Plato, Pascal, Gauss and other great thinkers of the past.

The teaching methods included in the textbook emphasize the power of visual representations of mathematical concepts, the use of physical materials (manipulatives) and support of digital technology – the major pillars of an experimental (and/or experiential) approach to mathematics. Manipulative tools include two-sided counters, pattern blocks, square tiles, linking cubes, geoboards, coins, and dice. Digital tools include an electronic spreadsheet, computational knowledge engine *Wolfram Alpha*, and dynamic geometry software the *Geometer's Sketchpad.* The textbook emphasizes connections of mathematics to real life as many mathematical concepts and procedures (e.g., the algorithm of long division) are true reflections on common sense and intuitive understanding, something that schoolchildren (and their future teachers alike) inherently (often unconsciously) possess. For example, when children cut across the grass rather than walk along the pavement to get to the school building faster, they demonstrate intuitive understanding of the triangle inequality: the sum of any two side lengths of a triangle is greater than the third side length. Likewise, in the author's experience, even first graders intuitively recognize that the distance from a point on the floor to the wall is measured along the perpendicular. The goal of schooling in K-6 mathematics is to turn the intuitive into the conceptual. This aspect of teaching mathematics is an integral part of the textbook's pedagogy. Another important pedagogical aspect of the textbook is the demonstration that mathematics content and methods of teaching are not in the relation of dichotomy but rather, the diversity of methods that teacher candidates have at their disposal stems from the knowledge of content. Ultimately, the very appreciation of this diversity by teacher candidates not only benefits struggling students, but it opens a window to the teaching of extended content to students with special interest in mathematics (including "mathematically proficient students," to use terminology of the Common Core State Standards [2010]).

One of the major challenges of teaching and learning that face mathematics educators and their students nowadays is the appropriate use of technology, both physical and digital. The use of manipulative and computational learning environments is among the special features of the textbook. In particular, an electronic spreadsheet is used as an instrument of demonstration, computation, and exploration. The tool is used both pragmatically (e.g., for creating a diagram and/or for computational problem solving) and epistemically (e.g., for exploring possible extensions of a problem). The epistemic use includes technology-enhanced problem posing. The context of geometry is supported by *The Geometer's Sketchpad.* Furthermore, the grade appropriate use of computational knowledge engine *Wolfram Alpha*, not commonly recognized as being helpful at the primary level, is highlighted in the textbook. Conceptually, using the combination of manipulatives, mathematics, and computing allows one to appreciate what it means to make mathematical connections.

Following is an example of an infrequently considered technology-enabled connection among topics in elementary mathematics. In Chapter 4, linking cubes are suggested in support of finding the number of decompositions of the number 10 (or like number in terms of size) in three integer addends arranged from the least to the greatest. This mathematical activity is put in the context of building triples of towers out of linking cubes to be arranged in the non-decreasing order of their heights. Then, in Chapter 11, this activity is revisited under the umbrella of permutations in order to find all ways of building triples of towers out of the given number of linking cubes having been arranged in all possible orders. Finally, in Chapter 12, the obtained tactile-theoretical results are verified in the context of the spreadsheet simulation of rolling three dice, say, 5,000 times and comparing theoretical and experimental probabilities of having the sum of 10 (or other sums not greater than 18) on the faces of three dice. An important aspect of this third part deals with realization that not all differently ordered decompositions of the given whole number in three like addends may be used in the context of dice rolling as the largest number of spots on the face of a die is six. Therefore, through connecting additive decompositions of whole numbers to rolling dice, a K-6 teacher candidate can appreciate one of the major ideas that relate mathematics to real life: mathematical models have to be consistent with practical applications. Such connectivity of theory and practice, being one of the important features of the textbook, is in line with many features of STEM education. Other previously infrequently considered topics include basic ideas of combinatorics as a counting technique that goes beyond hands-on experimentation with concrete materials and the pedagogy of using conceptual shortcuts as an early algebra trial-and-error argument informed by conceptual understanding.

The textbook consists of thirteen chapters and appendix. Chapter 1 titled *Teaching PreK-K Mathematics* introduces counting and measurement as motivation for the development of the concept of number. Conservation of number and counting skills are discussed. The notion of invariance as one of the big ideas of mathematics is introduced in the context of counting. Images are shown as genesis of mathematical formulations and sources for fostering metacognition. These include the sums of consecutive natural numbers, consecutive odd numbers, consecutive even numbers, and the sum of numbers within a gnomon. The importance of the idea of a unit for comparing cardinal numbers is emphasized. Classification of pattern blocks in terms of color, size and shape is presented. The symbolism of the AB-patterns is discussed through the lens of early algebra.

Chapter 2 titled *Teaching Operations in Grades 1 – 4* begins with the demonstration of how arithmetical operations stem from physical actions using simple real-life context. Starting from problems with small numbers to allow for a picture-based solution, it is the increase of the magnitude of numbers involved that requires an action to be formalized in the form of an operation. Base-ten system is introduced by showing how addition and subtraction on a place value chart can be carried out as a game with certain rules defined by the system. Multiplication is shown as repeated addition in context and its commutativity is demonstrated through recourse to geometry. Multiplication table, relations among its entries and emerging patterns they form are discussed through the lens of hands-on conceptual explorations. Problems leading to the operation of division are presented through two models: measurement and partition. Long division is presented as a reflection on common sense, something that most people possess without any knowledge of division as an operation. Several numeric examples are considered to provide an answer as to why multiplication and division have precedence over addition and subtraction. Ambiguous cases are presented to motivate the use of parenthesis as a remedy for ambiguity.

Chapter 3 titled *Conceptual Shortcuts* introduces the notion of a conceptual shortcut in mathematics education as a problem-solving strategy based on insight that makes solution of a problem less computationally involved. The power of conceptual shortcuts is demonstrated through their application to several arithmetical tasks, including the summation of consecutive integers used, according to a legend, by young Gauss. It is shown how systems of two linear equations in two variables can be solved through a conceptual shortcut based on divisibility properties of coefficients.

Chapter 4 titled *Decomposition of Integers into Like Addends* begins with using towers built from linking cubes and the W^4S (**w**e **w**rite **w**hat **w**e **s**ee) principle to discuss several ways of decomposing a whole number in two, three, and four like summands. Multiple tactile and visual problem-solving strategies have been discussed as their appreciation by a problem solver is an indicator of his/her metacognitive development. These strategies will be revisited in Chapter 11.

Chapter 5 titled *Activities with Addition and Multiplication Tables* begins with connecting entries of the tables to perimeter and area of rectangles to demonstrate how arithmetic stems from geometry. Using *Wolfram Alpha*, 4 × 4 addition and multiplication tables are generated and connected to 16 types of rectangles. Multiple strategies for the summation of numbers in the tables are discussed. Using simple spreadsheet programming, odd and even products are generated in the 10 × 10 multiplication table and their quantities have been computed. The results of computations have been explained conceptually towards the end of opening a window into algebra.

Chapter 6 titled *Using Technology in Posing and Solving Problems* suggests that technology-enabled problem posing fosters metacognitive skills and provides teacher candidates with research-like experience in mathematics and its grade-appropriate pedagogy. Towards this end, the notions of numeric coherence, pedagogic coherence and contextual coherence are introduced and exemplified as didactical tools of mathematical problem posing. The joint use of spreadsheets, *Wolfram Alpha* and the On-Line Encyclopedia of Integer Sequences (OEIS®) is demonstrated. Technology is shown as a tool for eliciting discovery rather than merely performing computations. It is argued that through technology-enabled discovery students can develop skills in asking conceptual questions. The context of cookies on plates is used to introduce Fibonacci numbers the knowledge of which, in turn, as shown in Chapter 13, inform spreadsheet modeling. It is shown how a combination of concrete objects and digital tools represents an engaging learning environment which motivates problem posing and supports problem solving.

Chapter 7 titled *Fractions* begins with the introduction of the part-whole and the dividend-divisor contexts for fractions as an extention of measurement and partition models for the division of whole numbers. The latter context makes it possible to inaugurate the idea of equivalence as one of the big ideas of mathematics not previously met through the study of integer arithmetic. The use of manipulatives in the context of the set model for fractions supports one's understanding of numeric equivalence in mathematics. Addition and subtraction of fractions is introduced contextually by reflecting on the same operations in whole number arithmetic. A rule of multiplying two fractions is developed using a real-life context and a rectangular grid. The meaning of the Invert and Multiply rule is explained in terms of the change of unit by demonstrating how it works in the case of integers and then extended to fractions using the measurement and the partition models for division. Representation of fractions on a number line is introduced in the dividend-divisor context for fractions through the process of fair division of rectangular pies. Unit fractions are treated as fractional analogues of multiples of ten and used to estimate proper fractions employing a dividend-divisor context, thus making numeric inequalities more friendly for the learners of mathematics. A brief history of Egyptian fractions is presented and the Greedy algorithm is connected to its use by *Wolfram Alpha* in finding an Egyptian representation of a proper fraction. The application of Egyptian fractions and the use of other means of dividing pizzas fairly are discussed, and three different partitioning methods are compared in terms of their effectiveness. A complexity of seemingly simple activities with pizzas is revealed through this comparison.

Chapter 8 titled *From Fractions to Decimals to Percent* begins with demonstrating how the decimal representation of a common fraction stems from the algorithm of long division. Terminating and non-terminating (periodic) decimals are considered. The use of *Wolfram Alpha* in generating decimal representations of fractions and the emerging complexity of mathematics behind the transition from fractions to decimals is discussed. The concept of percent is introduced through the comparison of (some) common fractions within a spreadsheet-based 100-cell grid. The grid is used to demonstrate where the rule of multiplying decimal fractions comes from. Several word problems with friendly numbers in different notations allowing for a pictorial solution are considered. Pictorial solutions serve as means of conceptualization needed for formal problem solving with arbitrary data. The importance of the unit in defining fractional parts is emphasized.

Chapter 9 titled *Ratio as a Tool of Comparison of Two Quantities* begins with using the contexts of measurement and partition for the introduction of the concept of ratio. Some classic examples of ratios are presented, including the appearance of the Golden Ratio in geometry and the manifestation of its invariance under variations made possible by dynamic geometry software. Another context for ratio is comparison of quantities in real-life situations. Classroom examples of incorrect use of proportional reasoning are considered and explained in terms of the notion of misconception. The concept of rate defined as a special ratio between two related quantities in different units is illustrated using real-life examples leading to the notion of the rate of change.

Chapter 10 titled *Geometry* begins with considering a scalene triangle and making a distinction between three points and three segments as entities of mathematics associated with this fundamental geometric figure. A hands-on activity with triangles and quadrilaterals and its replication in the context of the *Geometer's Sketchpad* aimed at informal measuring of the sums of the shapes' internal angles is, once again, connected to the notion of invariance. Different ways of dividing a rectangle in two identical parts are discussed towards the end of showing multiple geometric meanings of the fraction 1/2 and demonstrating how shapes with the same area may have different perimeters. Van Hiele levels of geometric thinking are introduced and activities with pattern blocks aimed at the transition from Level 0 to Level 1 are discussed and interpreted in terms of Vygotskian theory of child development. Finding areas of different shapes on a geoboard is used to demonstrate where Pick's formula comes from. Briefly, the Pythagorean triples are introduced through special cases of geoboard-based triangles with integer perimeters. The difference between informal and formal deduction is discussed. Tessellation with triangles and quadrilaterals is presented as hands-on activities extended to the context of dynamic geometry software. This extension makes it possible to demonstrate how a physical movement can be defined as a formal geometric operation. At the level of informal deduction, areas of rectangles sharing same perimeter and perimeters of rectangles sharing same area are computed for specific values of perimeter and area. It is explained why, given perimeter, no rectangle with the smallest area exists and why, given area, no rectangle with the largest perimeter exists. This explanation motivates the transition from area of rectangle as the number of unit squares it includes to area of rectangle as the product of its side lengths.

Chapter 11 titled *Elements of Combinatorics: Counting Through a System* starts with the introduction of a tree diagram as a tool conducive to counting through a system. Different types of trees are considered and explored in the contexts of the rule of product, the rule of sum, and the rule of sum of products. Permutation of objects is demonstrated using a tree diagram. By using sets with distinct and repeated objects, factorials are introduced as tools for computing permutations of objects in such sets. It is shown how permutation in the case of repeated objects gives a way to the concept of combination. Combinatorial ideas are used to count additive decompositions of whole numbers with reference to Chapter 4. These results (along with those of Chapter 4) will be used in Chapter 12 in the context of comparing experimental and theoretical probabilities using a spreadsheet. Using pattern blocks, combinations without and with repetition are introduced through two distinct ways of selecting objects from a set and describing those selections using multiple-letter words with exactly two different letters. This description allows one to use skills of counting permutations of letters in a word to develop formulas for computing the two kinds of combinations.

Chapter 12 titled *Elements of Probability Theory and Data Management* begins with emphasizing experimental character of situations with uncertain outcomes. A computer spreadsheet is introduced in the context of computationally exploring coin tossing and dice rolling. The notion of sample space is motivated by the need to measure chances numerically. Different representations of a sample space in the contexts of tossing coins and rolling dice are discussed. It is shown how chances can be compared on a pre-operational level using a specially designed spreadsheet representing chances in the form of bar graphs. Representation of data is demonstrated using a line plot, a histogram and a stem and leaf plot with support of Excel and *Wolfram Alpha*. Mean, median and mode as the basic measures of central tendency are introduced. Their limitations as statistical tools are discussed opening a window to more advanced concepts of data management.

Chapter 13 titled *Programming Details of Technology Used in the Textbook* includes the uses of the *Geometer's Sketchpad* in constructing fraction circles, equilateral triangles, squares, and regular polygons with more than four sides, in tessellating with triangles and quadrilaterals, as well as programming details of spreadsheets used throughout the textbook.

Finally, Appendix includes 300 problems and open-ended questions presented through 13 activity sets.

The textbook reflects the author's almost three decades of preparing teacher candidates, both at the undergraduate and graduate levels, to teach mathematics in the elementary classrooms of the United States and Canada. The university where the author works since 1998 is located in the United States in close proximity to Ontario province of Canada and many of the author's students are Canadians pursuing their master's degree in education. The textbook is written to be of interest to faculty members in the colleges of education and to those in mathematics departments who are assigned to teach a mathematics content and methods course for elementary teacher candidates.

In conclusion, with deep respect I wish to express my sincere gratitude to Rok Ting Tan for inviting me to submit a book proposal to World Scientific that, with greatly appreciated support and recommendations of anonymous reviewers, resulted in the present publication. Many thanks are due to Nijia Liu for invaluable editorial guidance during my final work on the book. Last but not least, I acknowledge being under obligation to Maria G. Freitas Ladouceur of SUNY Potsdam for critical help with numerous technology-related issues.

Sergei Abramovich
Potsdam, NY

Contents

Preface vii

Chapter 1 Teaching PreK-K Mathematics 1

1.1 Developing the concept of number 1
1.2 Learning to think with images 7
1.3 Classification 10
1.4 Visual patterns and their symbolic description 12

Chapter 2 Teaching Operations in Grades 1-4 15

2.1 From actions to operations 15
2.2 Addition and subtraction in base-ten system 18
2.3 Multiplication 20
2.4 Division 28
2.5 A note on the order of arithmetical operations 34

Chapter 3 Conceptual Shortcuts 37

3.1 Conceptual shortcuts in arithmetic calculations 37
3.2 Conceptual shortcuts in multiplying two-digit numbers 38
3.3 Conceptual shortcuts in the summation of integers 39
3.4 Conceptual shortcuts in solving word problems 41

Chapter 4 Decomposition of Integers Into Like Addends 45

4.1 Decomposition in two addends 45
4.2 Decomposition in three addends 47
4.3 Alternative decomposition strategies 50
4.4 Decomposition through creating spaces among addends 53

Chapter 5 Activities with Addition and Multiplication Tables 55

5.1 Geometrization of arithmetical tables 55
5.2 Counting all numbers in addition and multiplication tables 55

5.3 Counting numbers with special properties in multiplication and addition tables 64

Chapter 6 Using Technology in Posing and Solving Problems 69

6.1 Technology as a link between problem solving and problem posing 69
6.2 Didactical coherence in problem posing 70
6.3 From systematic reasoning to a joint use of the modern-day tools 72
6.4 From cookies on plates to spreadsheet modeling 77
6.5 Spreadsheets, diagrams and number lines 81
6.6 Posing new questions about cookies and plates 85
6.7 From cookies on plates to Fibonacci numbers 86
6.8 From cookies to creatures to equations 91

Chapter 7 Fractions 93

7.1 From unit fractions to two contexts for fractions 93
7.2 Adding and subtracting fractions 101
7.3 Multiplying two proper fractions 103
7.4 Multiplying two improper fractions 106
7.5 Dividing fractions 108
7.5.1 Dividing whole numbers 108
7.5.2 Dividing proper fractions 110
7.6 Conceptual meaning of the *Invert and Multiply* rule 111
7.6.1 The case of dividing whole numbers 112
7.6.2 The case of dividing fractions 113
7.7 Representing fractions on a number line 118
7.8 Unit fractions as benchmark fractions 120
7.9 Egyptian fractions with applications 123

Chapter 8 From Fractions to Decimals to Percent 131

8.1 From long division to fractions as decimals 131
8.2 From alternative comparison of fractions to percent 134
8.3 Word problems with fractions, decimals, and percent 137

Chapter 9 Ratio as a Tool of Comparison of Two Quantities **143**

9.1 Different definitions of ratio 143
9.2 Introducing ratio as a tool 145
9.3 Using ratio to find an unknown quantity 146
9.4 Problems that require insight to avoid an error in using proportional reasoning 149
9.5 Rate as a special ratio 151
9.5.1 Different problems involving rate 151
9.5.2 Solving the car and the truck problem using rectangular grids 153

Chapter 10 Geometry **157**

10.1 Basic concepts of geometry 157
10.1.1 Triangle 157
10.1.2 Rectangle 158
10.2 The van Hiele levels of geometric thinking 160
10.3 Basic activities at Level 1 of the van Hiele model 162
10.4 Geoboard activities and Pick's formula 165
10.5 Tessellation 171
10.5.1 Tessellation with scalene triangles 173
10.5.2 Tessellation with quadrilaterals 175
10.6 Creative thinking leading to special rectangles 176
10.7 On the relationship between perimeter and area of a rectangle 179

Chapter 11 Elements of Combinatorics: Counting Through a System **185**

11.1 Tree diagrams 185
11.2 From tree diagrams to permutations 188
11.3 Permutation of objects in the sets with distinct and repeated objects 191
11.4 Using permutations in counting additive decompositions of integers 192
11.5 Combinations without and with repetition 194

Chapter 12 Elements of Probability Theory and Data Management **199**

12.1 Teaching about chances 199
12.2 Randomness and sample space 201
12.3 Different representations of a sample space and Pascal's triangle 203
12.4 Understanding chances in a computer environment 207
12.5 Fractions as tools in measuring chances 209
12.6 Graphic representations of numeric data 212
12.7 Measures of central tendency 214

Chapter 13 Developing Technology Used in the Book **217**

13.1 Constructing fraction circles using the *Geometer's Sketchpad* (GSP) 217
13.2 Constructing an equilateral triangle using the GSP 219
13.3 Constructing a square using the GSP 220
13.4 Constructing a regular pentagon using the GSP 221
13.5 Tessellation with triangles and quadrilaterals using the GSP 222
13.5.1 Tessellation with triangles 222
13.5.2 Tessellation with quadrilaterals 224
13.6 Programming of the *Cookies on Plates* spreadsheet 225
13.7 Programming of the *Postage* spreadsheet 227
13.8 Programming of the spreadsheet of Fig. 5.10 229
13.9 Programming of the spreadsheet of Fig. 5.11 230
13.10 Programming of the spreadsheet of Fig. 5.12 231
13.11 Programming of the spreadsheet of Fig. 10.19 232
13.12 Programming of the *Tossing Four Coins* spreadsheet 233
13.13 Programming of the spreadsheet of Fig. 12.10 234
13.14 Programming of the M&M spreadsheet of Figures 12.7 and 12.8 235

Appendix — Activity Sets: 300 Problems and Questions **237**

Activity Set 1: Conceptual shortcuts, the order of operations, and arithmetic on place value charts 237

Activity Set 2: Connecting mathematical practice with mathematical content of additive decompositions of integers 241
Activity Set 3: Explorations with multiplication and addition tables 245
Activity Set 4: Using technology in solving and posing problems 248
Activity Set 5: Learning to move from visual to symbolic 251
Activity Set 6: Unit fractions as benchmark fractions 256
Activity Set 7: From pre-operational to operational level of dealing with fractions 258
Activity Set 8: Pizza as a context for learning fractions 262
Activity Set 9: Word problems with fractions, percentages, and ratios 263
Activity Set 10: Informal geometry 265
Activity Set 11: Combinatorics 267
Activity Set 12: Probability and data management 269
Activity Set 13: Open-ended questions 272

Bibliography 275
Index 283

CHAPTER 1: TEACHING PREK-K MATHEMATICS

1.1 Developing the concept of number

A number when presented as a symbol (whatever the notation) is an abstraction. For example, the number 3 (the Hindu-Arabic notation, introduced by Fibonacci[1] into Western European mathematics) is an abstraction. It becomes concrete entity through the association with concrete objects, like three apples, three cars, three cats, three birds, and so on (Fig. 1.1). In that way, the number 3 is a common decontextualized characteristic of the four groups of the above-mentioned (and many more not mentioned) objects. Contextualization gives birth to various concepts through a natural tendency of thinking with images. For example, whereas all four sets of three objects (Fig. 1.1) can be described by the same abstract symbol 3, they are all different: three identical apples, three different cars, three cats put in two groups and three birds two of which are the same. Thinking with images can be very intuitive, like recognizing (Fig. 1.1) that there are more ways of arranging the cars in different orders than arranging the birds in different orders. This intuitive recognition might result in a child's actual activity of ordering the cars and the birds on the level of images. A teacher must be prepared to accommodate such advanced mathematical thinking of young children by appreciating the dual power of context as a means for comprehending abstraction and as a milieu for the development of new ideas.

Fig. 1.1. Contextualizing the number 3.

[1] Leonardo Fibonacci (1270–1350, Italy), the most prominent mathematician of his time.

The most important role of context at the PK level (age 4-5) is that through the use of different sets of objects with the same cardinality the concept of number develops. Historically, numbers stem from tally marks used to count objects and to record the results of counting. Tally marks do not use the concept of place value which allows one to represent records of counting of a large number of objects more efficiently by arranging objects (or their representations through tally marks) in groups according to a certain rule. The base-ten system, commonly used in the modern civilization, has its genesis in the ancient Egyptian civilization dating back to the 4th millennium B.C. when, in the absence of positional notation, every power of ten was represented by a special sign [Katz, 2007; Rudman, 2007]. Much later, with the advent of Hindu-Arabic numerals into mathematics, the meaning of each digit in a base-ten number was made depended on its position within the number. For example, in the number 313 the meanings of the digit 3 are different for the first and the last digits. Indeed, the first and the last digits show, respectively, that when in the set of objects with cardinality 313, the objects are put in groups of hundreds, tens, and ones, there would be three groups of hundred objects and three groups of a single object in each. The importance of understanding of the meaning of digits in a base-ten number is reflected in notes and guidance for teachers in England expecting pupils at an early age to "become fluent and apply their knowledge of numbers to reason with, discuss and solve problems that emphasise the value of each digit in two-digit numbers" [Department for Education, 2013, p. 11].

Measurement is another real-life context for the use of numbers. This context was the primary source of expanding integers to rational numbers and then to irrational numbers. Unlike counting, measurement has various mathematical contexts. In geometry, one can measure lengths, areas, and volumes. This was probably the first context in which measurement was used as evidenced by the Egyptian papyrus roll (ca. 1650 B.C.) found in 1858 by Henry Rhind, a Scottish scholar and collector of antiques [Chace, Manning, and Archibald, 1927]. One of the problems recorded in the papyrus roll deals with determining the area of an isosceles triangle with a narrow base in which case a lateral side and the altitude to the base are almost indistinguishable in size (enabling the ancient mathematicians to compute areas of such triangles, as half the product of the base by the side, with a good precision without using what about a millennium later became known as Pythagorean theorem). The focus on an isosceles triangle in the Egyptian papyrus roll was due to the affinity to such shapes of ancient (10,2000 B.C. – 2,000 B.C.) architecture of Neolithic culture [Kuijt, 2002]. Approximately at the same time as the relationship between the side lengths of a right triangle became associated with the name of Pythagoras[2], the idea of integrating theoretical knowledge with practical applications was seen by Confucius[3] through social lens as the major role of education when teaching is an activity through which both teachers and students develop, and learning is a process inseparable from thinking which, in turn, implies learning [Dai and Cheung, 2015].

In arithmetic, using the base-ten positional system, measurement may involve counting objects in a set through arranging them into the clusters of ones, tens, hundreds, and so on with the maximum of nine identical items in each cluster. According to the Standards for Preparing Teachers of Mathematics in the United States [Association of Mathematics Teacher Educators, 2017, p. 49], in order to have deep understanding of early elementary mathematics, "teachers need to understand how counting relates to place value ... [so that] every counting number can be expressed in a unique way as a numeral made of a string of digits". In England, the main goal "of mathematics teaching in key stage 1 [age 5-7] is to ensure that pupils develop confidence and mental fluency with whole numbers, counting and place value ... supported by objects and pictorial representations" [Department for Education, 2013, pp. 5, 6]. For example, in order to count the (large) number of linking cubes in a box (Fig. 1.2, bottom), one can create a rod of ten cubes (by counting them in the rod) and then use the rod as a unit of measurement when creating new rods (this time without counting cubes), then count rods to have ten such rods (Fig. 1.2, top) that can be used as a new unit of measurement, and so on.

---

[2] Pythagoras of Samos (570 – 495 B.C.) – a Greek philosopher traditionally credited with many mathematical results, the Pythagorean theorem being the most famous one.

[3] Confucius (551 – 479 B.C.) – a Chinese philosopher traditionally credited with many educational ideas.

Likewise, other units of measurement can be used in the context of counting. Indeed, eggs can be measured in terms of egg cartons, light bulbs can be measured by packs, bananas can be measured by weight, and so on. But in each case, we have a specific unit of measurement, something that in the case of counting objects was represented by just a single object, regardless of its nature. That is, whereas in some cases (e.g., Fig. 1.2) measurement can supplant counting, the former activity is much more complex and consequential than the latter one. As mentioned by mathematics educators in Japan, "in measuring length of strings or weight of water, the quantity can be divided infinitely and cannot always be expressed in integers" [Takahashi, Watanabe, Yoshida, and McDougal, 2004, p. 71].

Fig. 1.2. Counting through measuring.

In order to count objects in a set correctly, counting skills have to be developed. The skills are based on the following mathematical knowledge, both conceptual and procedural. One has to know number names and the order of numbers. One has to use each number name only one time when counting and count each object only one time. One has to begin counting with the number one and know that the last number name used in the process of counting is the total number of objects counted. Furthermore, the order in which objects are counted does not affect the total count and objects can be rearranged to facilitate making one-to-one correspondence between the objects and number names (counting labels). In other words, the cardinality of a set of objects is invariant under different ways of counting objects in the set. This counting principle makes it possible to introduce the notion of invariance – " a property of a mathematical object which remains unchanged when the object undergoes some form of transformation ... (e.g. re-arrangement or manipulation)" [Ministry of Education Singapore, 2020, p. 8] – as one of the big ideas of the entire mathematics.

Research suggests that the basic principles of counting listed above have to be developed first in order to use counting as a skill [Gelman and Meck, 1983]. Although one can conceptualize the infinity of numbers (indeed, as long as we have a way of writing down integers as large as we want, every integer (alternatively, whole number, natural number) is followed by itself increased by one and, therefore, this process may never stop), in real life, the number of objects in a set is finite and thus the process of counting stops when the last object in the set is assigned a number name which is then called the cardinality of the set. That is, infinity, not having a clear frame of reference, belongs to much higher level of abstraction than a whole number.

Conservation of a number is another skill that has to be developed. To conserve a number requires one to understand that the number of objects does not change when the objects are covered, put in several groups, or just rearranged. This understanding is especially important when counting objects loosely organized because the efficiency of counting depends on the way objects are arranged. For example, counting a number of seats at a round table or even at a rectangular table is more difficult than counting the same number of lined-up seats (Fig. 1.3).

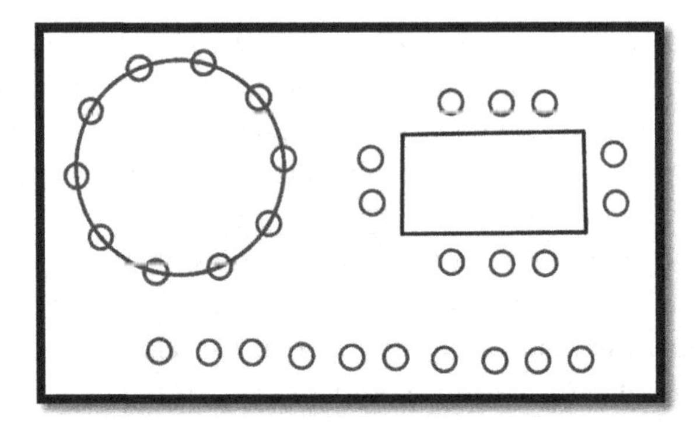

Fig. 1.3. Practicing skills to conserve a number.

Furthermore, a skill of conserving a number can be applied in the case of distribution of any material like pouring water from a bottle into several glasses and realizing that the glasses contain the same amount of water that was in the bottle. The same is true when a cake is cut into several pieces and put on several plates. The original quantity of the cake does not change through this process. Subitizing is yet another concept associated with counting – it is the ability to correctly determine a small number of objects without counting them [Kaufman, Lord, Reese and Wolkman, 1949]. Typically, human ability to subitize is limited to six objects.

The earliest distinction among numbers was based on the idea of even and odd numbers grounded in the experience of putting objects in pairs. In order to understand the idea, one has to be introduced to its genesis. History of mathematics tells us that the game of guessing odd or even with respect to the number of coins (or other small objects) held in hand was considered ancient even in the time of Plato[4]. The game, most likely, included pairing objects held in hand and, therefore, the concepts of even and odd numbers were associated with this action to decide the outcome of the game. Therefore, young children must be introduced to this experience in order to develop conceptual understanding of even and odd numbers. Appreciating differences that exist among numbers in a game situation is one of the first big ideas of mathematics and its pedagogy which opens the door for transition from acting on concrete objects to

[4] Plato (ca. 428-348 B.C.) – a major figure in the history of Western philosophy.

using culturally accepted mathematical notation. As stated by educators in Singapore, mathematics teaching focusing on "big ideas can help students develop a deeper and more robust understanding of mathematics and better appreciation of the discipline" [Ministry of Education Singapore, 2020, p. 19]. As a result, with deep understanding and appreciation of the applied character of mathematics, supported by careful guidance of their teachers, "students learn that they can apply the skills they acquire in mathematics to other contexts and subjects" [Ontario Ministry of Education, 2020, p. 67].

1.2 Learning to think with images

In ancient times, numbers and operations on numbers were associated with geometric images, the creation of which was the result of a certain action. Nowadays, recommendations for mathematics preparation of elementary teachers include the use of "drawings, diagrams, manipulative materials, and other tools to illuminate, discuss, and explain mathematical ideas and procedures" [Conference Board of the Mathematical Sciences, 2012, p. 33]. The discussion of mathematical ideas and procedures through the use of their images develops metacognition by providing students with an "opportunity for thinking about thinking in the classroom" [Ministry of Education Singapore, 2020, p. 21]. The genesis of using drawings and geometric patterns as representations of numbers can be found in ancient, ca. 1000 B.C., Chinese mathematics [Beiler, 1964] As Aristotle[5] put it, "the soul never thinks without an image" (cited in [Arnheim, 1969, p. 12]). For example, the (physically created) image shown in Fig. 1.4 was associated with the summation of consecutive whole numbers 1, $1+2$, $1+2+3$, $1+2+3+4$, ... ; and the corresponding sums were (and are) called triangular numbers due to the image they form. Likewise, the image shown in Fig. 1.5 was associated with the summation of consecutive odd numbers 1, $1+3$, $1+3+5$, $1+3+5+7$, ... , and those partial sums of odd numbers can be rearranged (through action) by using gnomons[6] (Fig. 1.6) to form squares, so that $1+3=4$, $1+3+5=9$, $1+3+5+7=16$,

---

[5] Aristotle (ca. 385-323 B.C.) – a Greek philosopher and polymath, student of Plato.

[6] Heron of Alexandria, a Greek mathematician and engineer of the 1st century A.D., defined a gnomon as something which, when added to or subtracted from a number or a shape, yields a similar formation.

and $4 = 2 + 2$, $9 = 3 + 3 + 3$, $16 = 4 + 4 + 4 + 4$. Figs 1.7 and 1.8 show how the summation of consecutive even numbers can be reduced to either the summation of consecutive whole numbers or the summation of consecutive odd numbers. Indeed, Fig. 1.7 shows that $2 + 4 + 6 + 8 = 2 \cdot (1 + 2 + 3 + 4)$ and Fig. 1.8 shows that $2 + 4 + 6 + 8 = (1 + 3 + 5 + 7) + 4$. One can see that "if the count is carried out correctly, the count will always be the same" [Ontario Ministry of Education, 2020, p. 115].

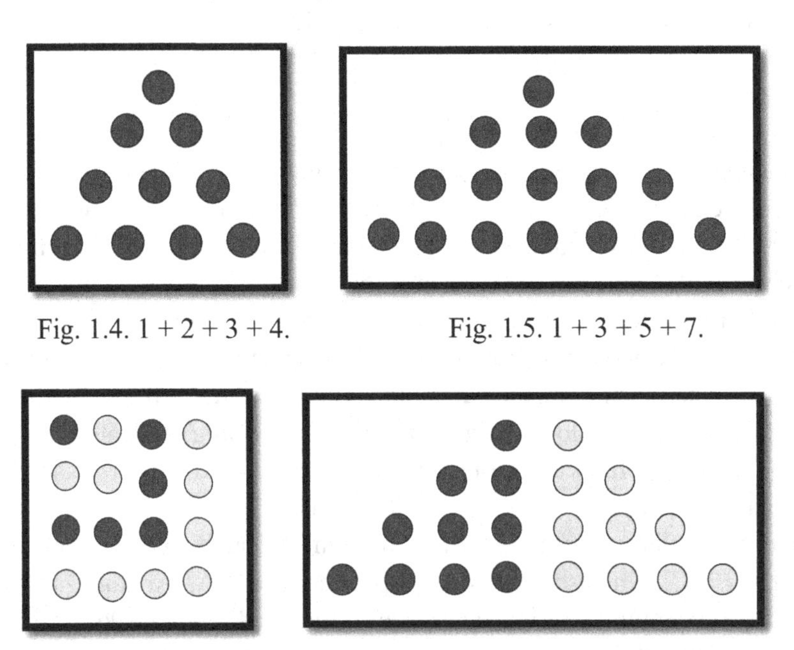

Fig. 1.4. $1 + 2 + 3 + 4$.

Fig. 1.5. $1 + 3 + 5 + 7$.

Fig. 1.6. $1 + 3 + 5 + 7 = 4 \times 4$. Fig. 1.7. $2 + 4 + 6 + 8 = 2 \cdot (1 + 2 + 3 + 4)$.

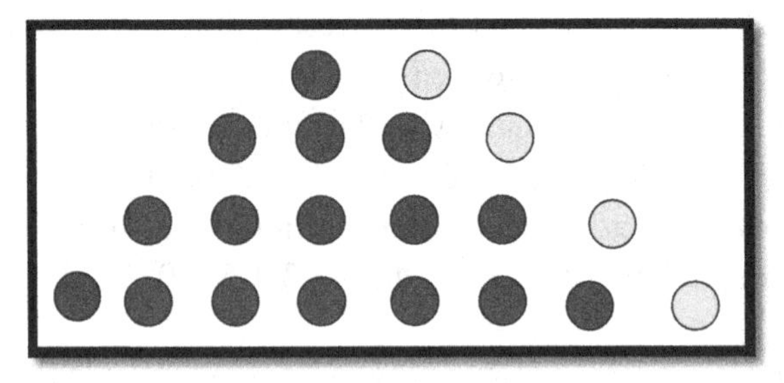

Fig. 1.8. $2 + 4 + 6 + 8 = (1 + 3 + 5 + 7) + 4$.

After the concept of (whole) number is developed, the next step is to explain that, in contextualized situations, numbers represent the results of counting or measuring. However, with young children's inability to conserve a number, often numbers have the spatial meaning rather than the numeric meaning. For example, by seeing on a picture three elephants and five mice (Fig. 1.9) or two groups of five identical mice, each group taking different space on a picture (not shown here), a child may not see that five is greater than three or that the number of mice in each group is the same. This may be seen as one of the earliest misconceptions in the study of mathematics regarding numeric meaning of symbols describing quantities appearing in one's visual field. Whereas counting skills do require knowledge of the order of numbers at the level of linguistic memorization, these skills do not necessarily include knowledge of numeric comparison of numbers used in counting. Therefore, numeric comparison relies on visualization which, without conceptual understanding may lead young learners astray (see also Chapter 8, Section 8.3, Fig. 8.6).

Towards the end of addressing this visual vs. numeric misconception, note that in order to compare the numbers 5 and 3 in terms of greater vs. smaller, they have to be presented through the same unit, like it is shown in Fig. 1.10. In order to see that the number of triangles is the same as the number of rectangles (Fig. 1.11), one has to establish one-to-one correspondence between the two types of shapes. This is how one learns the numeric order of numbers visually, without counting. In fact, mathematics educators in South Africa discourage students to use counting as an operation when dealing with large numbers and, instead, "move learners away from using tallies (which represent counts) when they operate on numbers" [Department of Basic Education, 2018, p. 33]. Large numbers are then compared through the operation of subtraction (or using base-ten system): $543 > 461$ because $543 - 461 > 0$ (or because there are more hundreds in 543 than in 461). Likewise, $543 > 537$ because (with the same number of hundreds) 543 has more tens than 537. Finally, $543 > 542$ because (with the same number of hundreds and tens) 543 has more ones than 542.

Fig. 1.9. Is 3 greater than 5?

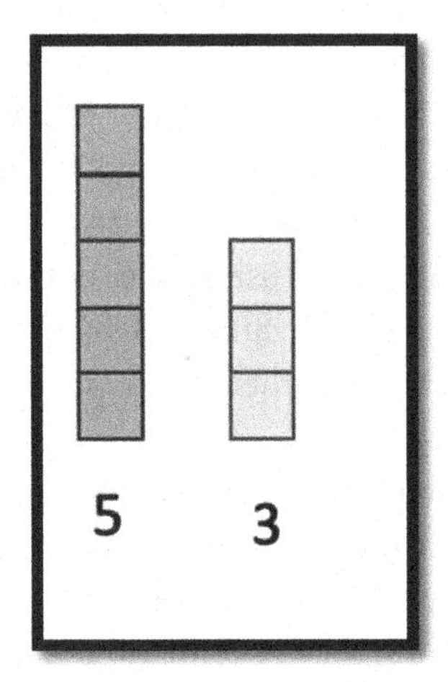

Fig. 1.10. Comparing 5 to 3 using same units.

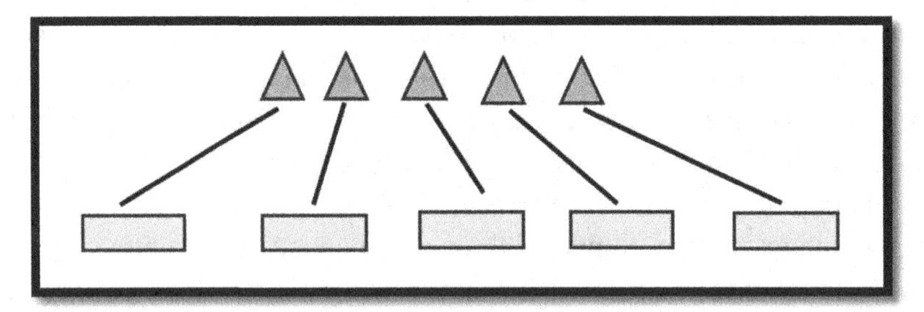

Fig. 1.11. Comparing cardinalities through one-to-one correspondence.

1.3 Classification

The presence of the basic geometric shapes in the mathematics classroom (called pattern blocks, see https://en.wikipedia.org/wiki/Pattern_Blocks) prompts the activity of their classification as a way of abstracting their common characteristics. Basic characteristics of geometric images (or pattern blocks) are shape, size, and color. Also, there may be the binary classification (stemming from yes – no or on – off dualities); e.g., triangles and not triangles, squares and not squares, and so on. As an aside, note that two-sided counters – the modern-day teaching tools of the elementary mathematics classroom – are physical embodiments of the binary system the genesis of which can be found in the 1st millennium B.C. Chinese mathematics and its first scientific uses were due to the writings of Leibniz[7] in the early 18th century.

Shapes can be put together to form other shapes. According to Common Core State Standards [2010, p. 9], kindergarten students are expected to "use basic shapes to construct more complex shapes". For example, one can make a rhombus, an isosceles trapezoid, and a (regular) hexagon out of two, three, and six equilateral triangles, respectively. Likewise, the hexagon can be made out of triangle, rhombus, and trapezoid. Such action-oriented relationships among pattern blocks are very important to be interpreted in terms of arithmetic for they lead to the concept of additive partition of the whole into unit fractions. Indeed, the latter relationship can be expressed through the equality $\frac{1}{6}+\frac{1}{3}+\frac{1}{2}=1$ which ascribes the unit fractions 1/6, 1/3, and 1/2 to the triangle, the rhombus, and the trapezoid, respectively, and the number 1 to the hexagon (the whole). Likewise, the triangle can be called a one-half of the rhombus, a one-third of the trapezoid, or a one-sixth of the hexagon (Fig. 1.12). That is, the same object can be assigned different fractional names depending on the unity (whole) of which the object is a part (alternatively, fraction). In particular, the fact that six equilateral triangles (or an isosceles trapezoid, a rhombus, and an equilateral triangle) form a hexagon is an elementary precursor of the concept of tessellation discussed in Chapter 10. When basic geometric and numeric ideas are taught in a holistic way (i.e., being integrated), children at an early age appreciate what they learn as "underlying mathematical concepts ... [and fathom] that skills are not taught as procedures only" [Ministry of Education Singapore, 2020, p. 21].

---

[7] Gottfried Wilhelm Leibniz (1646-1718) – the great German mathematician and philosopher.

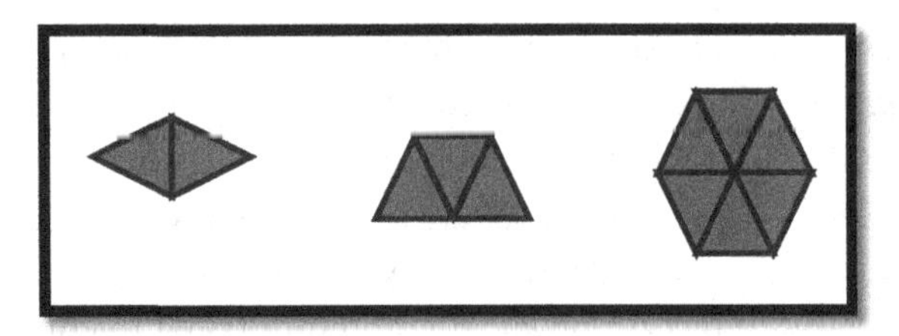

Fig. 1.12. Hexagon as a combination of other shapes.

1.4 Visual patterns and their symbolic description

Another major idea of PK-K mathematics curriculum is creating, recognizing and describing visual patterns. Symbolic description of visual patterns can be seen as early algebraic activities in the broad context of learning mathematics. The presence of a pattern implies repetition. For example, the pattern, in which the pair triangle-square is repeated as long as one wishes, can be described as an AB-pattern where the letter A represents a triangle, and the letter B represents a square (Fig. 1.13). A pattern, in which the quadruple triangle-triangle-square-square is repeated as long as one wishes, may be described as an AABB-pattern as well as an AB-pattern. In order to justify the latter description, one can introduce a pattern where hundred triangles are followed by hundred squares; that is, a set of two hundred shapes repeats itself over and over. Describing such pattern in the form $\underbrace{AA\ldots A}_{100\ times}\underbrace{BB\ldots B}_{100\ times}$ cannot be accepted because this description can hardly be pronounced, and such a large number of letters cannot be accurately repeated. So, here one can learn about the efficiency of symbolic representation, one of the earliest introductions to formal reasoning in mathematics. In other words, recognizing that multiple visual patterns have the same symbolic description may be referred to as an early algebraic skill of generalization.

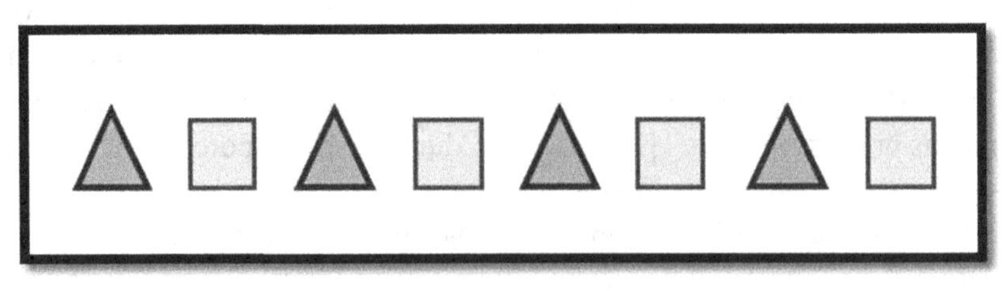

Fig. 1.13. From visual to symbolic: an AB-pattern.

Students can be creative and, when asked to design a pattern, often present designs following a pattern which cannot be described using the "AB" language. An example of such a pattern is shown in Fig. 1.14. In a slightly different form such pattern was suggested by one of the author's students, an elementary teacher candidate [Abramovich, 2020a]. One can see that towers built out of blocks are developing in the sets of three towers and each element of the second triple is twice as large as the corresponding element of the first triple, and each element of the third triple is three times as large as the corresponding element of the first triple. So, if the pattern continues in that way, each element in the fourth, fifth, sixth, and so on triple would be four, five, six, and so on as large as the corresponding element of the first triple. Therefore, the pattern can be described numerically through the sequence 1, 2, 3, 2, 4, 6, 3, 6, 9, 4, 8, 12, 5, 10, 15, and so on. A question that can be asked about this pattern is purely numeric one, yet having been connected to the visual representation in the form of the triples of towers: How tall would be the 20th tower in the sequence of the triples of towers? Such question, seeking information, can even be asked by a curious kindergarten student without realizing that an answer would require mathematical knowledge beyond that level. In their turn, teacher candidates have to realize that within a context which young children understand, questions seeking information yet requiring a much higher grade-appropriate explanation can often be asked.

According to Isaacs [1930], young children ask two types of questions: information-type questions and explanation-type questions. An art of teaching elementary mathematics is to address such questions in some way, even by physically building 20 towers following the pattern through measuring heights. Knowledge beyond K-level is required to answer this question mathematically (i.e., to provide an explanation). To this end, one has to find the largest number smaller than 20 which is divisible by three (or can be reached through counting by threes when starting from zero; that is, first adding three to zero). Such number is 18 which is a 6-multiple of 3 (six times counting by threes starting from zero leads to 18). The number 18 corresponds to the 18-th tower in the pattern. Therefore, the next tower is the first one in the 7th triple, implying that the

20th tower would be 14-block tall. Similarly, such information-type questions can be asked about towers of much larger ranks (e.g., 100) but only mathematically they can be answered. This may help young children appreciate the need to learn mathematics as a useful method of studying the world around them. Many more interesting questions about this pattern, that can motivate the discussion of various reasoning tools, will be asked in other chapters of this book.

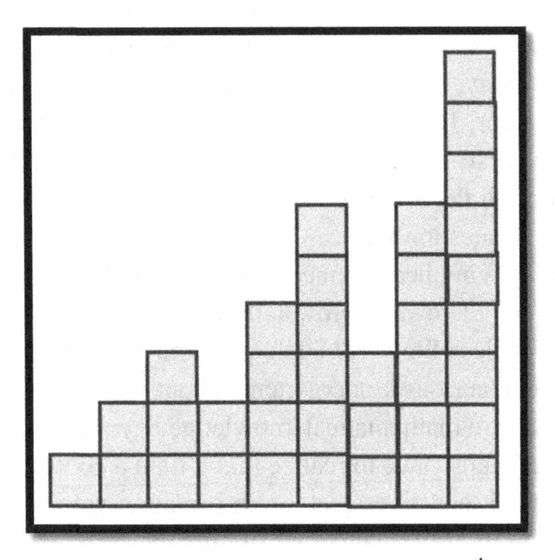

Fig. 1.14. How many blocks are in the 20th tower?

CHAPTER 2: TEACHING OPERATIONS IN GRADES 1-4

2.1 From actions to operations

What do students know before they begin the study of arithmetical operations, the first of which is addition? They know such action-oriented notions as "one more than a number" and "two more than a number", something that can be interpreted as augmenting a number by one and two, respectively. That is, adding one and two, respectively, to a number. For example, 6 is one more than 5 and 7 is two more than 5. Once the operation addition is formally introduced, knowing that six is one more than five can be interpreted as $5 + 1 = 6$ and knowing that seven is two more than five can be interpreted as $5 + 2 = 7$. Immediately, the two arithmetical facts have to be committed to memory (including many similar addition facts). Moreover, students know doubles of some one-digit numbers. For example, doubles of 1, 2, 3, and 4 are, respectively, 2, 4, 6, and 8. The knowledge of doubles can later be interpreted as multiplying the number by two. In the context of addition, the doubles can be interpreted as the sums of two equal numbers: $1 + 1 = 2$, $2 + 2 = 4$, $3 + 3 = 6$, and $4 + 4 = 8$, and so on. Furthermore, doubles always represent even numbers[8]. As educators in England suggest, such mathematical activities develop students' "recognition of patterns in the number system [for example, odd and even numbers), including varied and frequent practice through increasingly complex questions" [Department for Education, 2013, p. 6]. The recognition of a hidden complexity in a question asked by a student, whatever their grade level, develops through experience in answering questions and seeing them as a means that drives learning. That is why, in Chile, a future teacher is expected to know "how to deal with difficult mathematical questions posed by students and how to use them to generate

---

[8] A curious child may ask as to why this is true (an explanation-type question). Indeed, assuming that an even number represents the cardinality of a set of objects all of which can be put in pairs, a child might see a difference between the doubles of 3 and 4. However, in the spirit of early algebra (Chapter 1, Section 1.4), one can see the number 3 as a new unit (a single) which, when doubled, yields a pair. Likewise, one can see the number 4 as a new unit so that two units form a pair. Other ways of answering the child's question are possible. Teachers have to keep in mind that what one considers being obvious, is not in fact obvious, and, thereby, might require a sophisticated explanation.

learning outcomes" [Felmer et al., 2014, p. 11]. However, all these notions are only pre-requisites for learning addition as an operation, learning that can be motivated by the need to solve problems.

In teaching operations, one has to begin with contextual problem solving, when an operation is not needed to solve a problem because the numbers used by a context are small and simple counting, physical manipulation and just common sense may be enough to get an answer. For example, *if there are two cookies on Jonny's plate and three cookies on Annie's plate, how many cookies are there on both plates*?

Fig. 2.1. Counting vs adding.

Fig. 2.2. Counting vs subtracting.

A first grader can draw the pictures of two and three cookies (Fig. 2.1) and find through counting that there are five cookies on both plates. Likewise (Fig. 2.2), if out of six apples, four are small, one can count the remaining apples and conclude that there are two large apples. In other words, *if among six apples four apples are small, how many large apples are there*? The last question may be characterized as a whole-part (not to confuse with part-whole context for fractions, Chapter 7, Section 7.1) type subtraction problem. Alternatively, this is a take away type problem, when out of six apples four small apples were eaten and one has to find out how many apples remain. One can also introduce comparison context for subtraction: *if Jonny has 5 baseball cards and Ronny has 9, how many more cards does Ronny have than Jonny*? Another context is completion: *if Annie's homework incudes ten problems and she has solved six of them, how many problems does she need to solve in order to complete the homework*? All the problems can be easily solved, and one might wonder why we need to learn something called addition and subtraction. The answer deals with the need to solve similar contextual problems with large numbers. Young children are familiar with large numbers and know how to count beyond 100. However, finding the number of apples in two baskets with 78 and 183 apples or comparing the quantities of apples in the baskets through drawing and counting, are boring and time-consuming activities. This boring aspect of mathematical problem solving, that forces students to continue using strategies of drawing and counting when numbers involved are not small, can serve as a motivation for the introduction of new concepts (operations) which are called addition and subtraction.

So, the problem with cookies can be solved by using the former operation and having the solution in the form $2 + 3 = 5$ where each term in the left-hand side of the last addition fact is equal to the number of cookies on each plate. Likewise, the solution to the problem with 6 and 2 apples can be presented through the latter operation in the form $6 - 2 = 4$. In both contexts, the numbers 2 and 3 as well as 6 and 2 have situational referents in the form of cookies and apples, respectively. Knowing that $2 + 3 = 5$ (through counting), one has to be encouraged to write the sum as $3 + 2$ seeing it as two more than the number 3, and then find out that the "new" sum is also 5. In doing so, "students understand and use ... previously established results in constructing arguments" [Common Core State Standards, 2010, p. 6]. Contextualization confirms that $2 + 3 = 3 + 2$; that is, the order of plates through which one begins adding cookies does not change the sum. This is called commutative property of addition (the term may be omitted in the first-grade classroom; instead, one can talk about property of swapping the numbers/addends). Understanding this property, whereas not requiring knowledge of complicated terminology, enables one to recognize the following: knowing that, say, 25 is a number which is two more that 23 makes it possible to find the sum 2 + 23 by replacing it with the sum 23 + 2. At the same time, the difference 6 – 2 is not equal to the difference 2 – 6 as this operation, when contextualized, indicates impossibility of eating more apples than there are available. In other words, as one cannot swap 6 and 2, subtraction is not a commutative operation.

2.2 Addition and subtraction in base-ten system

The next idea regarding addition and subtraction is the use of base-ten system. The reason for using base-ten system appears to be purely biological – humans have ten fingers. Yet base ten is as difficult for young children as any other base different from base ten for elementary teacher candidates who used to do arithmetic in base ten alone. Therefore, one should not take for granted that the addition fact 5 + 6 = 11 is an obvious one. Intuitively, it is not obvious unless one understands that the meanings of the two ones in the right-hand side of the last equality are different. For that, one has to begin using place value charts and learn doing addition with the help of such charts (Fig. 2.3). The use of the charts can be introduced as a game with the following rule: one may not have more than nine objects within the same section (place value) of the chart where addition process is carried out. If there are more than nine objects in the section for ones, then the double nine is the largest number that can be greater than nine after two numbers are added (9 + 9 = 18). In that case, one has to create ten ones, turn them into one ten and move it to the section for tens. After adding two two-digit numbers, the largest number of tens in the section for tens is nineteen (99 + 99 = 198). In that case, one has to create ten tens, turn them into one hundred – a new place value. So, adding two-digit numbers requires a chart with at most three sections: for ones, for tens, and for hundreds.

Subtraction can also be demonstrated using place value charts when subtraction is an action that can be described as taking away some quantity from a larger (or equal) quantity. But taking away on a place value chart (Fig. 2.4) is another game with its own rules when ones are taken away until it is possible and what remains to be subtracted has to be engaged with tens (by breaking one ten in ten ones), then tens with hundreds, and so on. The Standards for Preparing Teachers of Mathematics in the United States [Association of Mathematics Teacher Educators, 2017, p. 82, italics in the original] recommend using "the verb *regroup* rather than *borrow*" in the context of subtraction meaning, perhaps, that the verb borrow may trigger the idea of the need to return, something that, obviously, is not the case when the result of subtraction is not a negative number. At the same time, the word regrouping is used when one replaces ten ones with one ten (Fig. 2.3).

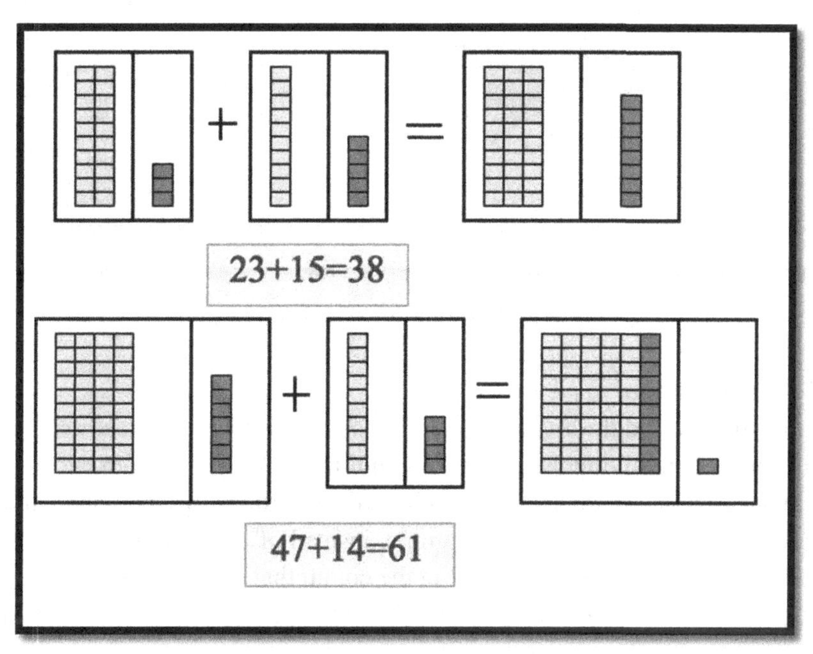

Fig. 2.3. Addition without and with regrouping.

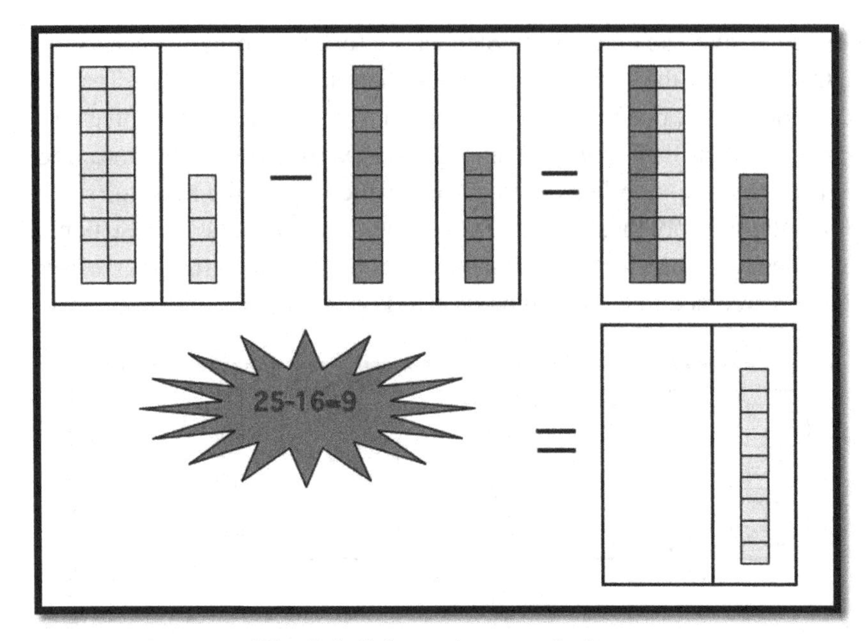

Fig. 2.4. Subtraction as coloring.

2.3 Multiplication

If operations of addition and subtraction stemmed from the need to deal with contexts involving large numbers, multiplication stems from the need to avoid dealing with a large number of repetitions of the same, not necessarily large, number. For example, instead of counting by twos 100 times, including the first number 2 in the count, that is, instead of carrying out addition $\underbrace{2+2\ldots+2}_{100\ times}$, the sum can be presented through what is called the product, 100×2, of two numbers (factors), 100 and 2, where the first number (multiplier) shows how many times the second number (multiplicand) is repeated. However, just as in the case of addition (or subtraction), multiplication is introduced in the context of problem solving when the number of (additive) repetitions of a small number is also small. To this end, consider the following problem/question: *How many cookies are there in five boxes, each box having six cookies*? The number of cookies in five boxes (Fig. 2.5) can be counted by writing down an addition sentence $6 + 6 + 6 + 6 + 6 = 30$.

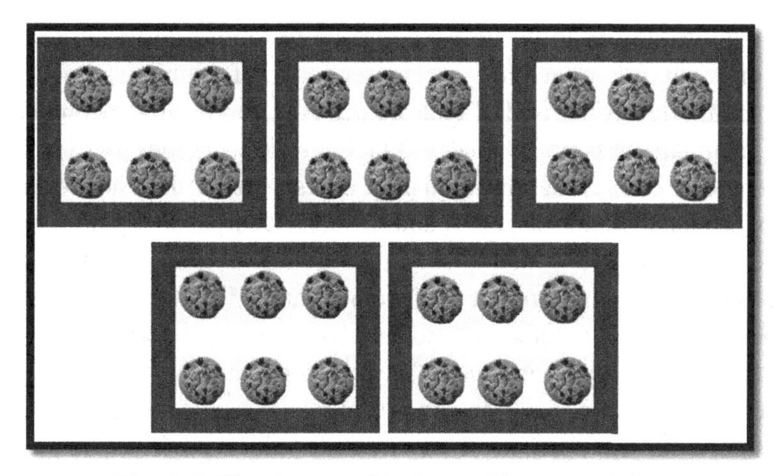

Fig. 2.5. Five boxes with six cookies in each box.

Asking the class to read this sentence loudly and to say how many *times* the number six was repeated, yields the answer: *five times six equals thirty*. The last statement can then be written in the form $5 \times 6 = 30$ in which a new symbol, "×", appears. This symbol, which, as noted in [Association of Mathematics Teacher Educators, 2017, p. 82], is not to be confused with the letter x used as notation for a variable, can then be called the multiplication sign. In turn, the new operation, connecting the numbers 5 and 6, is called multiplication. Many situations can be described through multiplication: counting the number of legs among five birds ($5 \times 2 = 10$), the number of wheels on three cars ($3 \times 4 = 12$), or the number of sides among four pentagons ($4 \times 5 = 20$). Note that while all the three cases do not require the use of multiplication, a large number of birds, cars, or pentagons would make repeated addition a time-consuming process requiring one to carry out repeated addition of a large number of addends. Therefore, one of the important pedagogic goals of mathematical teacher preparation in the United States is that "beginning Pre-K to Grade 2 mathematics teachers recognize the relationship between the content that precedes addition and subtraction (e.g., counting and cardinality) and the content that follows (e.g., multiplication and division)" [*ibid*, p. 51].

The next step towards developing multiplication skills is to show that multiplication is commutative, that is,

$5 \times 6 = 6 \times 5$; $5 \times 2 = 2 \times 5$; $3 \times 4 = 4 \times 3$; $4 \times 5 = 5 \times 4$; and so on. To this end, one can use a geometric representation of multiplication as a rectangular array comprised of unit squares (square tiles in the context of using manipulatives for a visual and tactile representation) as shown in Fig. 2.6. One can see that the number of unit squares within the rectangle can be counted both as 3×7 and 7×3. This shows how a more complicated operation, multiplication, requires a more sophisticated demonstration of the property of commutativity than addition (i.e., geometrization vs. counting). Also, this is an example when one can generalize from a single instance; that is, recognizing that $3 \times 7 = 7 \times 3$ implies $a \times b = b \times a$ for any two positive integers a and b. This is consistent with the position of mathematics educators in Japan where "students are expected to understand that an algebraic expression and a diagram may represent the same thing" [Takahashi et al., 2004, p. 195]. Note that the equality $5 \times 6 = 6 \times 5$ can be justified by taking out a cookie from each of the five boxes (Fig. 2.5) and creating a new box with five cookies so that the number of cookies in six boxes can be written as 6×5. Likewise, the equality $3 \times 6 = 6 \times 3$ can be justified by taking out three cookies from each of the three boxes and creating three new boxes with three cookies so that the number of cookies in six boxes can be written as 6×3. However, such demonstrations do not allow for a convincing generalization from the above two special cases: when factors differ by one and when the larger factor is a multiple of the smaller factor.

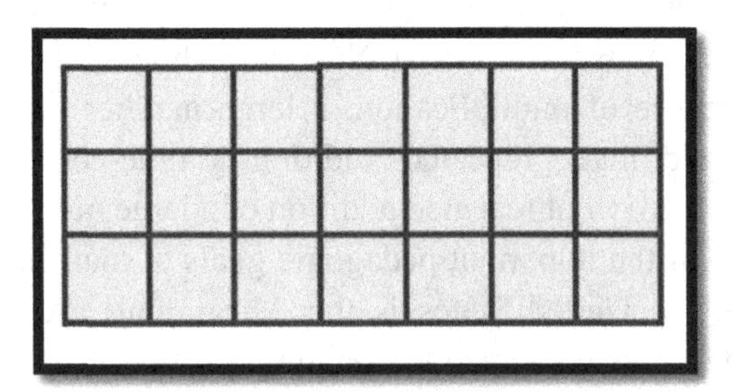

Fig. 2.6. Multiplication is commutative.

The next activity deals with the multiplication table – a collection of multiplication facts for small numbers. Consider the case of the multiplication table of size ten (Fig. 2.7). It is educationally beneficial to have students fill the table with the products of two integers by using conceptual understanding of multiplication as repeated addition. This approach to the teaching the multiplication table is emphasized by Japanese mathematics educators who stress the importance "to have students comprehend it by composing the table for themselves as they discover patterns in the table" [Takahashi et al., 2004, p. 136].

×	1	2	3	4	5	6	7	8	9	10
1	1	2	3	4	5	6	7	8	9	10
2	2	4	6	8	10	12	14	16	18	20
3	3	6	9	12	15	18	21	24	27	30
4	4	8	12	16	20	24	28	32	36	40
5	5	10	15	20	25	30	35	40	45	50
6	6	12	18	24	30	36	42	48	54	60
7	7	14	21	28	35	42	49	56	63	70
8	8	16	24	32	40	48	56	64	72	80
9	9	18	27	36	45	54	63	72	81	90
10	10	20	30	40	50	60	70	80	90	100

Fig. 2.7. The multiplication table with commutativity as symmetry.

The first row of the table develops from the number 1 through counting by ones. The second row of the table develops from the number 2 through counting by twos. The third row of the table develops from the number 3 through counting by threes. The fourth row of the table can be developed either through counting by fours beginning from the number 4 or by doubling the numbers in the second row. The doubling phenomenon is demonstrated in Fig. 2.8 in the case of $24 = 6 \times 4 = 2 \times (6 \times 2) = 2 \times 12$. The fifth row of the table develops from the number 5 through counting by fives. The sixth row of the table can be developed either through counting by sixes beginning from the number 6 or by doubling the numbers in the third row. The doubling phenomenon is demonstrated in Fig. 2.9 in the case of $24 = 4 \times 6 = 2 \times (4 \times 3) = 2 \times 12$. The seventh row of the table develops from the number 7 through counting by sevens.

The eighth row of the table can be developed either through counting by eights beginning from the number 8 or by doubling the numbers in the fourth row. The ninth row of the table can be developed either through counting by nines beginning from the number 9 or by tripling the numbers in the third row.

Alternatively, finger multiplication can be introduced. For example, in order to find the product 6×9, one bends the sixth finger counting from the left on both hands and reads the first and the last digits of the product as, respectively, the number of fingers to the left and to the right of the bended finger. This rule suggests that the sum of digits in the product of any integer not greater than 10 multiplied by 9 is always equal to 9 (as out of ten fingers, one finger is bended). This observation can be confirmed numerically as $6 \times 9 = 54$ and $5 + 4 = 9$. In order to explain why this rule works for any multiple of 9 in the multiplication table of size 10, let $9c = 10a + b$, $1 \le c \le 10$, $1 \le a \le 9$, $0 \le b \le 9$. Then $9c = 9a + a + b$ whence $a + b$ is divisible by 9 as both $9a$ and $9c$ are divisible by 9. Furthermore, $a + b = 9$. Indeed, from the identity $9c = 10\,(c - 1) + 9 - (c - 1)$ it follows that $a = c - 1$ and $b = 9 - (c - 1)$ whence $a + b = c - 1 + 9 - (c - 1) = 9$. One can see that in order to find a – the first digit of the product $9c$, one has to decrease c by one; something, that is expressed by the number of fingers to the left of the bended finger. Likewise, in order to find b – the second digit of the product 9c, one has to decrease 9 by $c - 1$; something, that is expressed by the number of fingers to the right of the bended finger. Finally, the tenth row of the table develops from the number 10 through counting by tens.

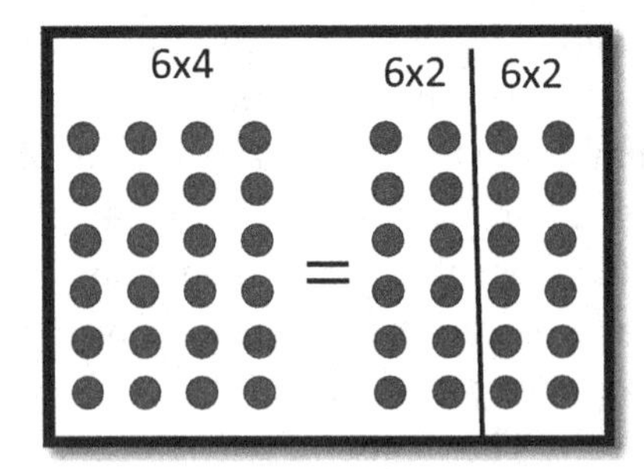

Fig. 2.8. Using a picture to explain a doubling phenomenon in the multiplication table.

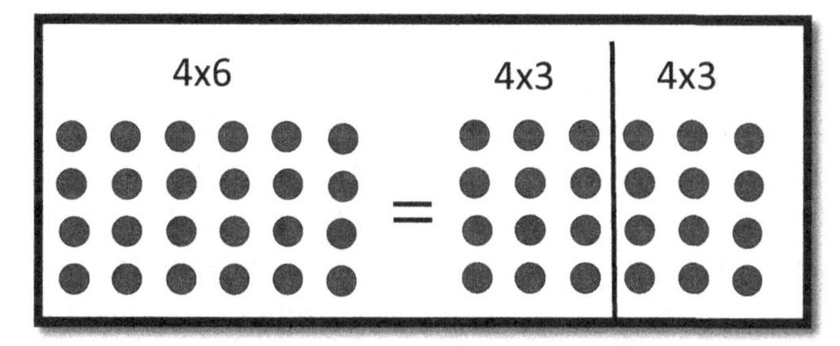

Fig. 2.9. Another picture-based doubling phenomenon in the multiplication table.

The multiplication table offers many opportunities for conceptual explorations using hands-on activities. Consider the table shown in Fig. 2.7. If such table is provided in hard copy with the main (top-left/ bottom-right) diagonal displayed, one can be asked to fold the (paper) table along this diagonal with numbers looking face-up and then punch a hole at any number on one side to see which number is affected on the other side. In Fig. 2.7 two tens are highlighted in the table. The ten on the right-hand side represents 5×2, the ten on the left-hand side represents 2×5. The same can be said about the number 24 (as well as about many other numbers). In that way, the commutative property of multiplication can be revisited through a hands-on activity.

Also, one can be asked to add numbers equidistant from the borders of the table within a column (or row). As shown in Fig. 2.10, the sums $3 + 30$, $6 + 27$, $9 + 24$, and so on, are the same. How can this be explained? In terms of multiplication, we have

$$3 = 1 \times 3, 30 = 10 \times 3, 3 + 30 = 1 \times 3 + 10 \times 3 = 3 \times (1 + 10) = 3 \times 11 = 33,$$
$$6 = 2 \times 3, 27 = 9 \times 3, 6 + 27 = 2 \times 3 + 9 \times 3 = 3 \times (2 + 9) = 3 \times 11 = 33,$$
$$9 = 3 \times 3, 24 = 8 \times 3, 9 + 24 = 3 \times 3 + 8 \times 3 = 3 \times (3 + 8) = 3 \times 11 = 33.$$

In terms of action, this phenomenon can be explained using images of the entries in the multiplication table in the form of rectangular arrays. Fig. 2.11 shows how moving a set of three counters from a larger array of counters exemplifies the replacement of a pair of numbers equidistant from the top and the bottom borders of the table by another such pair, so that the total number of counters in both arrays stays the same. One can note that the sums are not only all the same but are multiples of the number 11 which is one greater that the size of the table. Just as a table of size 10 is associated with the number 11, a table of size 9 can be associated with the number 10, a table of size 8 can be associated with the number 9, and so on. In general, in the table of size n, the numbers kl and $n - kl + 1$, $k, l = 1, 2, ..., n$, are equidistant from the top and the bottom borders of the table, and $kl + (n - kl + 1) = n + 1$.

×	1	2	3	4	5	6	7	8	9	10
1	1	2	3	4	5	6	7	8	9	10
2	2	4	6	8	10	12	14	16	18	20
3	3	6	9	12	15	18	21	24	27	30
4	4	8	12	16	20	24	28	32	36	40
5	5	10	15	20	25	30	35	40	45	50
6	6	12	18	24	30	36	42	48	54	60
7	7	14	21	28	35	42	49	56	63	70
8	8	16	24	32	40	48	56	64	72	80
9	9	18	27	36	45	54	63	72	81	90
10	10	20	30	40	50	60	70	80	90	100

Fig. 2.10. Adding numbers equidistant from the borders of the multiplication table.

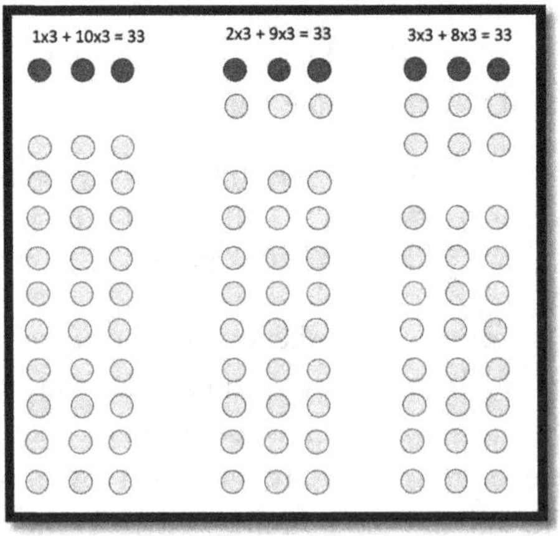

Fig. 2.11. A picture-based interpretation of a pattern in the multiplication table.

Consider the number 50 in the last row of the table of Fig. 2.7. If, starting from this number, one moves up-right diagonally, the numbers on such a path are 54, 56, 56, 54, and 50. One can see a symmetrical pattern when the path starts with 50 and ends with 50 by going up and down. In order to explain this pattern, consider the arrays shown in Fig. 2.12. The far-left array represents the number $50 = 5 \times 10$. The next array represents the number $50 = 6 \times 9$; that is, out of the array 5×10 one has to create the array 6×9. To this end, the 10th row with five circles is used in order to complement with a circle each of the five rows of five circles, and one needs another four circles to have a 6×9 array. This explains that the difference between 54 and 50 is 4. In order to make a transition from the array 6×9 to the array 7×8, two more circles are needed. Then the array 7×8 is replaced by the array 8×7 and this explains the presence of two consecutive 56's on the path in question. Then from the array 8×7 one moves to the array 9×6 and then to the array 10×5.

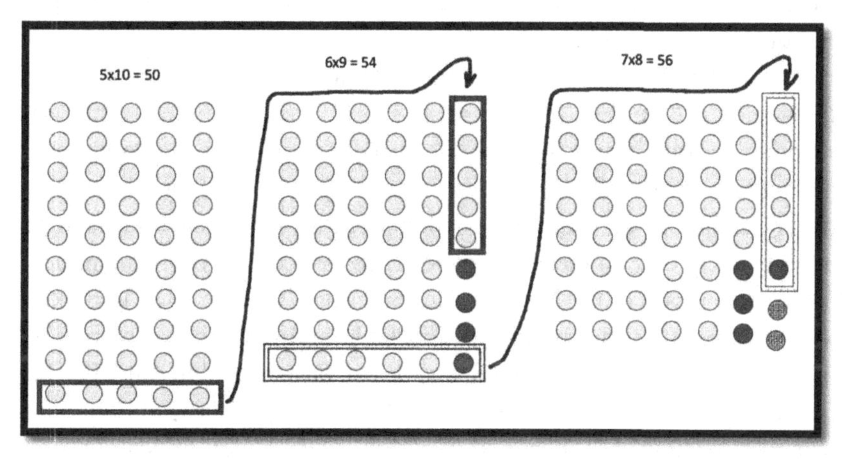

Fig. 2.12. Rearranging counters to explain a numeric pattern in the multiplication table.

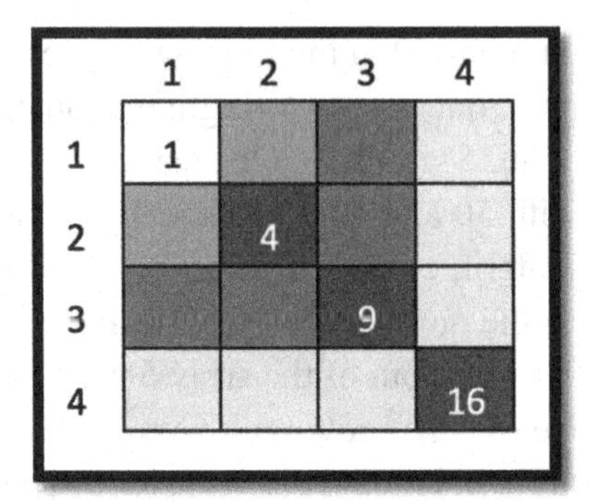

Fig. 2.13. Moving along the main diagonal by adding a double plus one.

Another activity deals with the sequence of perfect squares that the multiplication table contains. To this end, one can write down the numbers residing in the main diagonal of the multiplication table: 1, 4, 9, 16, and note that the differences between two consecutive numbers in this sequence are consecutive odd numbers. This observation can be explained by noting that in order to build a square of size $n + 1$ from the square of size n, one has to cover two adjacent sides with n unit squares and then add one unit square (Fig. 2.13, cases $n = 1, 2, 3$). In a different form, this phenomenon was demonstrated in Fig. 1.6 of Chapter 1 using gnomons. The multiplication table provides many opportunities for elementary teacher candidates to "study the mathematics they will teach in depth ... [rather than] to rely on their past experiences as learners of mathematics" [Conference Board of the Mathematics Sciences, 2012, p. 23], "to make a connection between the multiplication table and experiential activities" [Takahashi et al., 2004, p. 135] and learn how to guide students "to explore, investigate and find answers on their own" [Ministry of Education Singapore, 2020, p. 17]. For more activities with the multiplication table see Abramovich [2007, 2016a].

2.4 Division

Why do we need division as a new arithmetical operation? To answer this question, one can say that there are problems with a missing factor when the known numbers involved are large and such problems require a special mathematical operation called division. For example, *how can one put 781 apples into 11 boxes evenly*? This distribution of apples requires finding a number which when repeated 11 times gives 781. That is, one has to solve the equation $11 \times x = 781$. Alternatively, due to the commutative property of multiplication the last equation can be re-written as $x \times 11 = 781$ to mean that one has to find the number of repetitions of 11 needed to reach 781 (or the number which is smaller than 781 by less than 11). But in order to develop a skill in carrying out the described processes, the teaching of division must begin with problems, involving small numbers, that can be solved through a hands-on approach. To this end, two problems can be considered.

Problem 2.1. *Billy has 12 apples. He wants to give all the apples to his friends, Alan, Bob, and Tom, in a fair way. How many apples would each friend get?*

Fig. 2.14. Dividing 12 apples among 3 people.

Problem 2.2. *Billy has 12 cookies. He wants to make servings, 3 cookies in each serving. How many servings can he make?*

One can see (Fig. 2.14) how 12 apples are gradually partitioned among three people so that each person received four apples. In terms of multiplication, Fig. 2.14 represents the multiplication fact $3 \times 4 = 12$ where the second factor (the number of apples given to each of the three individuals) was not originally known and was found through a hands-on activity. In other words, Problem 2.1 can be solved as a multiplication problem with a missing second factor (the repeated number).

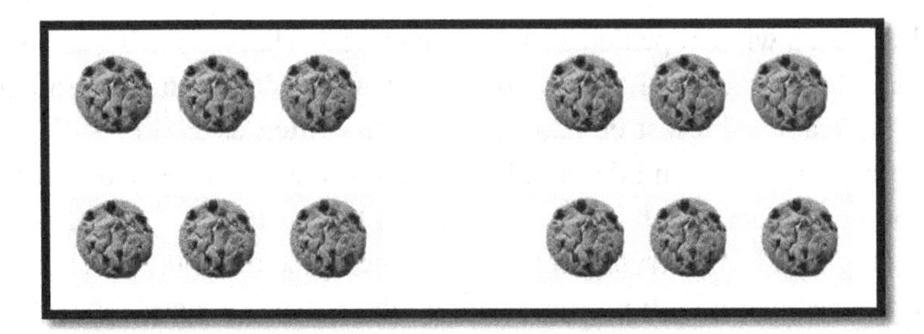

Fig. 2.15. Measuring 12 cookies by a 3-cookie serving.

Likewise, one can see (Fig. 2.15) how 12 cookies were used to create four servings of three cookies in each. In terms of multiplication, Fig. 2.15 represents the multiplication fact $4 \times 3 = 12$ where the first factor was not originally known and was found through a hands-on activity as well. In other words, Problem 2.2 can be solved as a multiplication problem with a missing first factor (the number of terms in the repeated sum).

Didactically speaking, Problem 2.1 introduces the partition model for division and Problem 2.2 introduces the measurement model for division. Both terms, partition and measurement, stem from actions that were used to solve the problems. When relating division to multiplication, "it is important to clarify which of the two values is being sought, the one corresponding to the multiplier or the one corresponding to the multiplicand" [Takahashi et al., 2004, p. 167].

In order to understand that division is not a commutative operation, one has to compare $12 \div 3$ to $3 \div 12$ by putting the two divisions in contexts of Problem 2.1 and Problem 2.2. One cannot measure 3 cookies by 12 cookies (Problem 2.2) and cannot divide 3 apples among 12 people (Problem 2.1) ... unless apples are cut in equal pieces, something that motivates the introduction of fractions in order to serve contexts where integers do not work. Even allowing for apples to be cut into equal pieces, dividing 12 apples among 3 people gives a different quantitative (in terms of apples) result than dividing 3 apples among 12 people and, in that sense, such comparison demonstrates the absence of commutativity for division. Using the context of Problem 2.2, one can show that whereas division by zero is not possible, the operation leads to the concept of infinity (already discussed in Chapter 1, Section 1.1). Indeed, out of 12 (or any number of) cookies one can create as many empty servings as one wants.

The multiplication table can be used to solve both partition and measurement division problems for relatively small numbers. For example, in order to find the result of dividing 56 by 7, one can use the table and find a number which when multiplied by 7 (i.e., repeated seven times) yields 56. However, beyond the multiplication table shown in Fig. 2.7, like in the case of $781 \div 11$, one needs to use a special tool known as the algorithm of long division (which, as will be shown in Chapter 8, does not just replace a calculator but, better still, the algorithm provides a number of conceptual clarifications in the context of fractions).

In order to explain how the long division works, a teacher has to demonstrate that each step of the algorithm can be put in a real-life context of division which requires common sense only; in other words, to make sure that a student is able "to justify, in a way appropriate to the student's mathematical maturity ... where a mathematical rule comes from" [Common Core State Standards, 2010, p. 4]. To this end, consider the following situation: *A school district received 462 oranges which have to be divided evenly among three cafeterias located in elementary, middle, and high schools.* The oranges arrived in boxes as follows (Fig. 2.16): there were four large boxes with the label 100, six small boxes with the label 10 and two individually wrapped up oranges. In order to resolve the situation, a manager, without any knowledge of mathematics, set aside three large boxes (each of which to be sent to the corresponding school) and only after that unpacked the fourth large box which included ten small boxes with the label 10 on each. The next step was to deal with 16 small boxes and put them evenly on the top of each of the three large boxes. This resulted in five small boxes to be sent to each of the three schools. Finally, the remaining box was unpacked, and 12 oranges were evenly divided among the three schools. As a result, each cafeteria received 154 oranges.

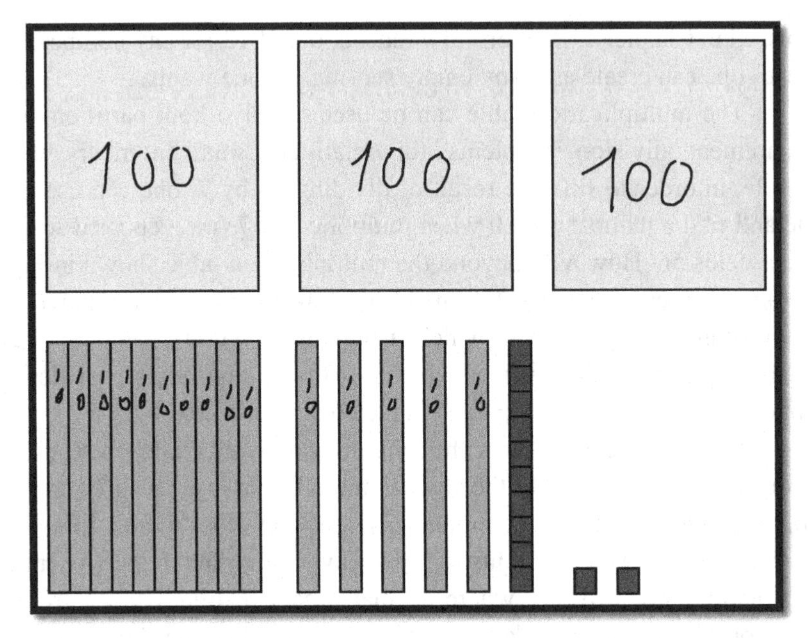

Fig. 2.16. Dividing 462 oranges among three cafeterias evenly.

The contextual description of the solution of the problem with the distribution of oranges can be presented in the decontextualized form shown in Fig. 2.17, in which 3 is divided into 462 by first dividing 3 into 4 (large boxes). The result, one box (for each school), becomes the first digit of the quotient. The next step is to subtract 3 (large boxes) from 4 (large boxes) to have 1 (large box). Now, one has to unpack this remaining box which reveals 10 small boxes and bringing down 6 exemplifies having the total of 16 small boxes which are to be evenly divided among the three schools. Dividing 3 into 16 (small boxes) results in the number 5 – the second digit of the quotient (contextually, five small boxes having been sent to each school along with a large box). Subtracting 15 (small boxes) from 16 (small boxes) yields a single small box which, when unpacked, reveals 10 oranges, thus leaving to the manager to partition 12 oranges among the three schools. That is, dividing 3 into 12 yields the number 4 – the last digit of the quotient. This completes the process of division without producing a remainder. Numerically, the result can be written in two alternative forms: $462 \div 3 = 154$ or $3 \times 154 = 462$. This is an example of what it means to "bring two complementary abilities to bear on problems involving quantitative relationships: the ability to *decontextualize* ... and the ability to *contextualize*" [Common Core State Standards, 2010, p. 6, italics in the original].

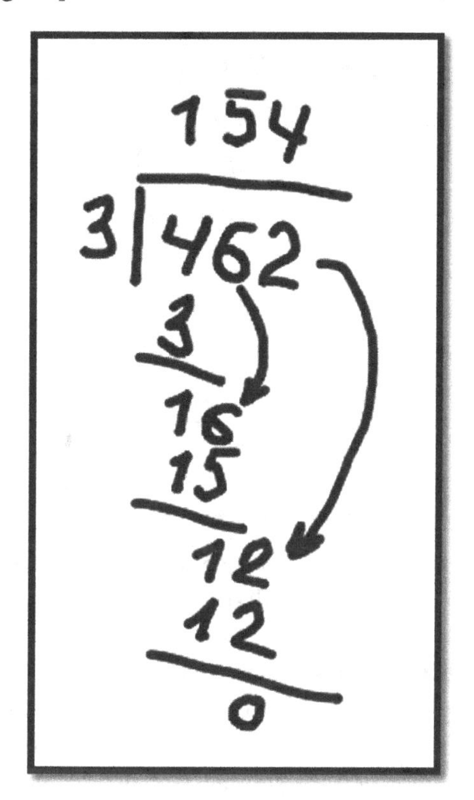

Fig. 2.17. From context (manager's task) to decontextualization (long division).

Finally, the original problem of putting 781 apples into 11 boxes evenly can be decontextualized allowing for the direct application of the algorithm of long division (Fig. 2.18). One can note that in both cases, division did not produce a remainder. Furthermore, an important observation is that remainder may not be greater than or equal to divisor. This observation would play an important role in the case of developing decimal representations of common fractions (Chapter 8).

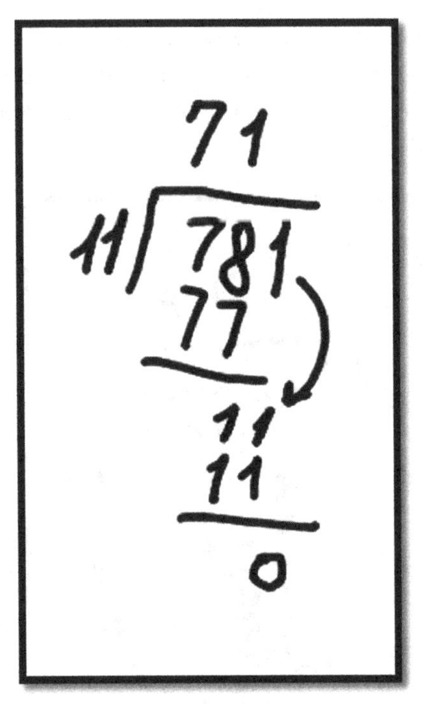

Fig. 2.18. A problem with 781 apples solved through the algorithm of long division.

2.5 A note on the order of arithmetical operations

Once an elementary teacher candidate raised a question about explaining the order of operations that students have to follow. Why do we multiply (or divide) first and then add (or subtract) ? This is a very good question as we often take the order of operations for granted. It looks like the order of operations is a convention: multiplication and division have to be carried out before addition and subtraction. The situation is even more ambiguous when multiplication and division are the only operations to be carried out. Teacher candidates need to know how to answer students' questions seeking explanation regarding the order of operations.

Consider the following example: Find the value of the numerical expression $5 \times 3 + 2$. Several cases may be considered if the order of operations is not defined. In one case, multiplication is performed first to have 15 and then the number 2 is added to the product to get 17. In another case, addition is performed first to have $3 + 2 = 5$ and then the sum 5 is multiplied by 5 to get 25. Yet, one can use the commutative property of addition to replace $5 \times 3 + 2$ by $2 \times 5 + 3$ and then perform addition first to have $2 + 5 = 7$ and then multiply 7 by 3 to get 21. Finally, commutative property of addition might prompt replacing $5 \times 3 + 2$ by $5 \times 2 + 3$, do the multiplication first to get 10 and then add 3 to get 13. Among the four answers, 17, 25, 21, 13 – which one is the right answer? One possible answer to this question is to note that multiplication is defined through addition rather than other way around and this distinction may suggest reducing everything to addition first and then complete the addition. Thus, we have $5 \times 3 + 2 = 3 + 3 + 3 + 3 + 3 + 2 = 17$. Using commutative property of multiplication does not change the result: $5 \times 3 + 2 = 3 \times 5 + 2 = 5 + 5 + 5 + 2 = 17$. The same justification can be applied to the case of $5 \times 3 - 2$ to have 13 as the answer. In other words, reduction to the same operation (before doing any operation) may be a way to clarify which answer is correct.

Consider the case of division and addition. Find the value of the numeric expression $12 \div 4 + 2$. Once again, several cases may be considered if the order of operations is not defined. First, one can begin with division, $12 \div 4 + 3$, and then do addition, $3 + 2 = 5$. Second, one can begin with addition, $4 + 2 = 6$ and then use 6 as divisor of 12 to get 2. Third, one can use commutative property of addition, $12 \div 4 + 2 = 2 + 12 \div 4$ and then do addition $2 + 12 = 14$ followed by dividing 14 by 4 to get 7/2. Finally, one can use commutative property of addition to replace $12 \div 4 + 2$ by $12 \div 2 + 4$, then divide 12 by 2 to get 6 and add 4 to get 10. Which of the four answers is correct: 5, 2, 7/2, 10? To answer this question, note that because division cannot be reduced to addition, one can replace the addend, 2, with division having the same divisor as 12. Thus, we have $12 \div 4 + 2 = 12 \div 4 + 8 \div 4 = \frac{12}{4} + \frac{8}{4} + \frac{20}{4} = 5$. So, 5 is the right answer.

There are cases when any order of operations yields the same answer. For example, consider the expression $8 \div 2 - 6$. Doing division first yields 4. Then $4 - 6 = -2$. Doing subtraction first yields $2 - 6 = -4$.

Then $8 \div (-4) = -2$. Whereas both answers are correct, the process of arriving to this answer, –2, is correct in the first case only.

Another ambiguous case can be represented by the expression $16 \div 2 \div 4$. Which division has one to do first? Dividing 16 by 2 first, yields 8 followed by $8 \div 4 = 2$. Dividing 2 by 4 first, yields 1/2 followed by $16 \div \frac{1}{2} = 32$. The rule here – a convention – to do division in the order of its appearance from left to right (i.e., to consider division as a left-associative operation). Another ambiguous case is when both multiplication and division are involved; for example, $16 \div 8 \times 2$. Starting with division yields $16 \div 8 = 2$ followed by multiplication $2 \times 2 = 4$ Starting with multiplication yields $8 \times 2 = 16$ followed by $16 \div 16 = 1$. Also, using commutative property of multiplication leads to $16 \div 8 \times 2 = 16 \div 2 \times 8 = 8 \times 8 = 64$. Once again, the rule here – a convention – to do operations in the order of their appearance from left to right. Note that in the case of addition and subtraction only, operations may be carried out in any order. For example, $14 + 5 - 3 = 14 - 3 + 5 = 16$. Such sequential unambiguity of addition and subtraction is due to their contextual meaning: in a perfect world, one can earn first and spend later as well as spend first and earn later. This observation will be used in the next chapter to introduce the notion of conceptual shortcut.

To conclude this section, note that the advent of computers gave birth to programming languages and the use of parentheses was designed to move away from ambiguity and convention. For example, in the numeric expression $(16 \div 2) \div 4$ the parentheses are used to indicate that the division $16 \div 2 = 8$ has to be completed first followed by $8 \div 4 = 2$. Likewise, $(16 \div 8) \times 2 = 2 \times 2 = 4$, $(5 \times 3) + 2 = 17$, $(12 \div 4) + 2 = 5$.

CHAPTER 3: CONCEPTUAL SHORTCUTS

3.1 Conceptual shortcuts in arithmetic calculations

Nowadays, mathematical problem solving is the blend of digital computing and formal reasoning. But even in the age of technology, reasoning is often considered to be more preferable strategy of doing mathematics than computing. Conceptual shortcuts [Canobi, 2005; Kuo, Hull, Gupta, Elby, 2013] in mathematics education can be defined as problem-solving strategies based on insight that makes solution of a problem less computationally involved. Yet insight, alternatively, productive thinking [Wertheimer, 1959], is difficult to formalize as a computational algorithm. To illustrate, consider the case of the combination of addition and subtraction with three whole numbers involved: $26 + 9 - 7$. Without much thinking, by carrying out operations in the left to right order permitted by the addition vs. subtraction sequential unambiguity (Chapter 2, Section 2.5), one can first do $26 + 9$, something that requires the use of addition with regrouping (Chapter 2, Section 2.2) by decomposing 9 into 4 and 5, so that $26 + 9 = (26 + 4) + 5 = 30 + 5 = 35$. The next step is to subtract 7 from 35 which requires one to 'borrow' two from one of the three tens included into 35 as follows: $35 - 7 = (35 - 5) - 2 = 30 - 2 = 20 + (10 - 2) = 20 + 8 = 28$. However, a conceptual shortcut is to begin with subtracting 7 from 9 to get 2 and then add 2 to 26 to get 28 – the same result but much less computationally involved, as the conceptual shortcut avoided changing the number of tens twice, i.e., first moving from two tens to three tens and then back, from three tens to two tens.

Of course, not all cases of addition and subtraction allow for the use of the conceptual shortcut of that kind by a young learner of mathematics. For example, in the 2nd grade, the expression $26 + 5 - 9$ cannot be computed without decomposing and borrowing/regrouping. So, a conceptual shortcut would be to decompose 9 into the sum of 5 and 4 followed by two subtractions: first subtract 5 from 5 to get zero and then subtract 4 from 26 to get 22; while both subtractions did not require knowledge of arithmetic in base ten, additive decomposition of 9 was required, something that was avoided in the previous example. Note that 2nd grade students do not yet have knowledge to replace $5 - 9$ by $-(9 - 5)$, so that $26 + 5 - 9 = 26 - (9 - 5) = 26 - 4 = 22$, as this conceptual shortcut requires knowledge of the 6th grade mathematics curriculum.

3.2 Conceptual shortcuts in multiplying two-digit numbers

Another example of a conceptual shortcut could be in multiplying 54 by 12 using a geometric representation of the product as shown in Fig. 3.1. The product can be represented by the sum of areas of four rectangles into which any integer-sided rectangle can be decomposed: $54 \cdot 12 = 50 \cdot 10 + 50 \cdot 2 + 4 \cdot 10 + 4 \cdot 2 = 500 + 100 + 40 + 8 = 648$, (assuming that students know that area of rectangle with whole number sides is the number of unit squares it includes – see Fig. 3.2 below and Fig. 2.6 of Chapter 2). This interpretation of area of a rectangle is described by Japanese mathematics educators as follows: "if you think about the meaning of the area, you can see that you can tile the entire area by unit squares and count the number of squares" [Takahashi et al., 2004, p. 221].

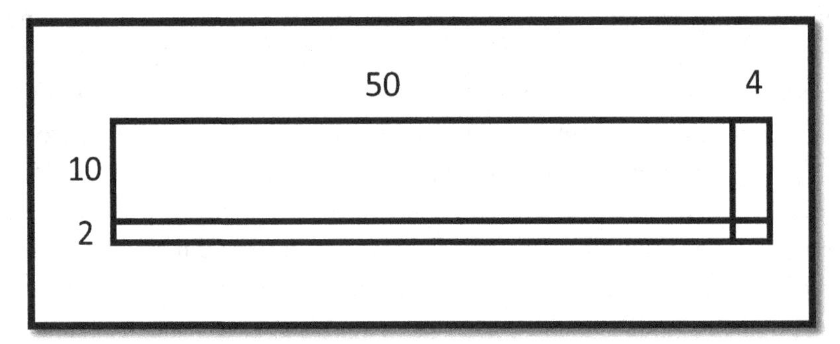

Fig. 3.1. Distributive property of multiplication over addition: $(50 + 4)(10 + 2)$.

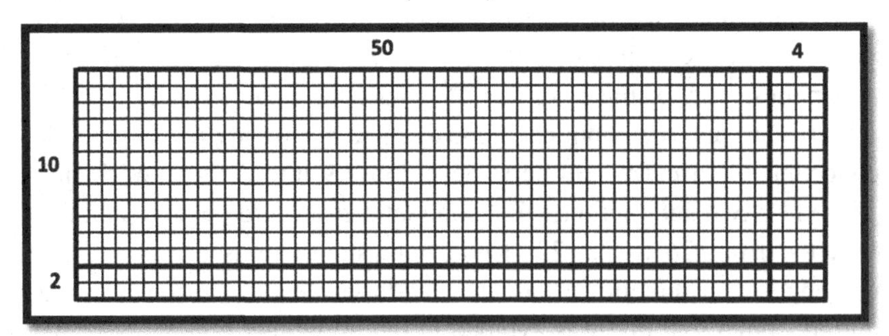

Fig. 3.2. Area of an integer-sided rectangle as the number of unit squares it includes.

There are several ways to multiply numbers without using a calculator based on the distributive property of multiplication over addition and the visual method is one of them. As mentioned by the Conference Board of the Mathematical Sciences [2012, p. 33] – an umbrella organization comprised of 19 professional societies in the United States concerned, in particular, with mathematics teacher education, teacher preparation programs should help teachers "to develop mathematical habits of mind ... [by using] mathematical drawings, diagrams, manipulative materials, and other tools to illuminate, discuss, and explain mathematical ideas and procedures." Similarly, mathematics educators in South Africa "want learners to move on from counting and calculating to seeing and using the structure of numbers and relationships between them" [Department of Basic Education, 2018, p. 26].

3.3 Conceptual shortcuts in the summation of integers

Conceptual shortcuts can be used in the context of summation of a large number of consecutive integers (or those in arithmetic progression when the difference between any two neighboring integers is the same; e.g., consecutive odd numbers 1, 3, 5, 7, ... are in arithmetic progression with the difference two). The following example is well known from the history of mathematics, the importance of knowing which, along with the social contexts, "affect [mathematics] teaching and learning" [Association of Mathematics Teacher Educators, 2017, p. 107]. A similar emphasis on the importance of knowing the history of mathematics by a teacher can be found in the primary mathematics standards in Chile as the "subject matter competency the Numbers Strand" [Felmer et al., 2014, pp. 10, 11]. The first activity with a number sequence is to find the sum of its consecutive terms starting from the first one. The following classic example deals with the summation of consecutive integers that can be described in terms of a conceptual shortcut.

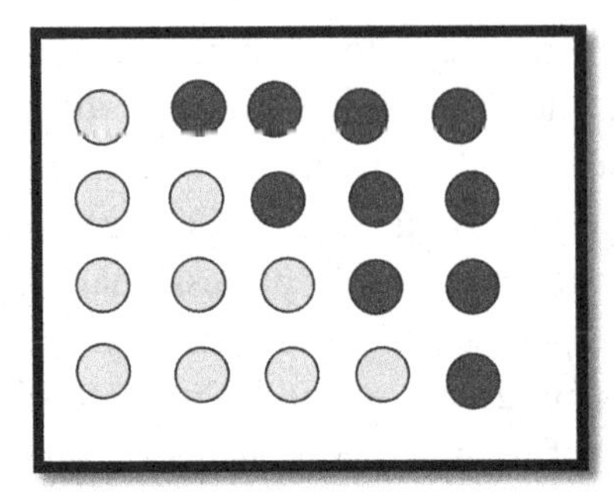

Fig. 3.3. An image of $1+2+3+4=\frac{4\cdot(4+1)}{2}$.

Once upon a time, a teacher asked his young students to find the sum 1 + 2 + 3 + ... + 100. Among the students was Gauss[9] who noted, as the legend goes, that if the sum is written backwards and added to the sum with the terms written forward, then the terms equidistant from the beginning and the end of the sum when added together yield the number 101. Indeed, 1 + 100 = 2 + 99 = 3 + 98 = ... = 50 + 51 = 101. There are hundred such pairs if the sum is computed twice. Therefore, the result has to be divided by two. In that way, $1+2+3\ldots+100=\frac{101\cdot100}{2}$. In general,

$1+2+3\ldots+n=\frac{n\cdot(n+1)}{2}$, as

$(1 + 2 + 3 + \ldots + k + \ldots + n) + n + (n - 1) + (n - 2) + \ldots + (n - k + 1) + \ldots$
$+ 2 + 1 = (1 + n) + [2 + (n - 1)] + [3 + (n - 2)] + \ldots + [k + (\mathrm{n} - \mathrm{k} + 1)] + \ldots$
$+ (n + 1) = n(n + 1)$.

The case $n = 4$ is shown in Fig. 3.3.

This conceptual shortcut can be applied to other arithmetic sequences; for example, the sum of the first four terms of the sequence 2, 4, 6, 8, 10, ... is presented visually in Fig. 3.4 from which it follows that $2+4+6+8=\frac{4\cdot(2+8)}{2}=20$. In general, the sum of n consecutive terms of the arithmetic sequence with the first term k and the difference d can be

---

[9] Carl Friedrich Gauss (1777-1855, Germany), commonly considered the greatest mathematician of all time.

found as follows:

$$k+(k+d)+(k+2d)+\ldots+[k+(n-1)d]=\frac{n[2k+(n-1)d]}{2}.$$

One can plug in the last formula $n = 4$, $k = d = 2$ to get 20. A different summation strategy based on a conceptual shortcut was shown in Chapter 2, Figs 1.5 and 1.6, for the sum 1 + 3 + 5 + 7 that allowed one to recognize a square number in the sum of the first four odd numbers through the formation of gnomons, another big idea from the history of mathematics.

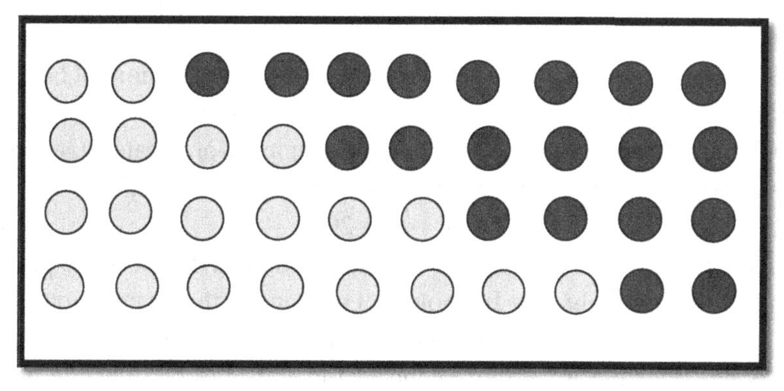

Fig. 3.4. An image of $2+4+6+8=\frac{4\cdot(2+8)}{2}=20$.

3.4 Conceptual shortcuts in solving word problems

Conceptual shortcuts can be used to solve algebraic word problems without recourse to manipulating unknowns through formal rules. Consider the following problem: *If Jim spent \$20 at a book sale to buy at least one of each of the books priced \$5 and \$2, how many books of each price did Jim buy?* To solve the problem, note that if Jim buys one, two, three or four books priced \$5, then the cost will be \$5, \$10, \$15, or \$20, respectively, and what would be left for buying \$2 books are \$15, \$10, \$5, or nothing. Only \$10 can be spent on \$2 books to buy five of them. Thus, Jim can only buy two books priced \$5 and five books priced \$2. This solution is shown in the diagram of Fig. 3.5.

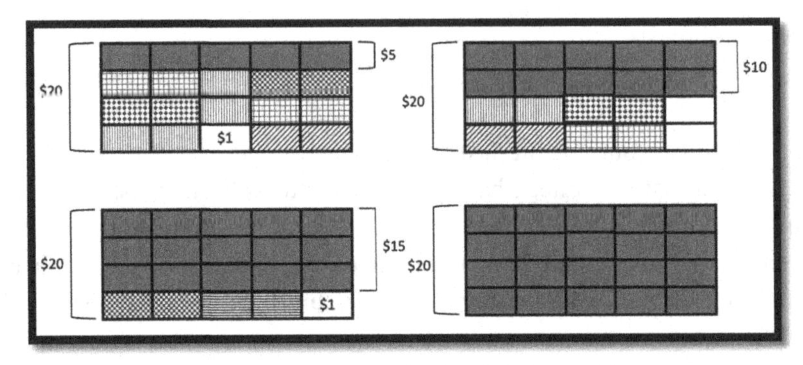

Fig. 3.5. Solution to the book sale problem.

Whereas one may not see the use of a conceptual shortcut here, at a formal level, when students in Grade 6 "extend previous understanding of arithmetic to algebraic expressions" [Common Core State Standards, 2010, p. 43], the problem can be described through the equation

$$5x + 2y = 20, \tag{3.1}$$

where the unknowns $x \neq 0$ and $y \neq 0$ stand for the number of books priced \$5 and \$2, respectively, that Jim bought. One can see that in equation (3.1) both $5x$ and 20 are divisible by 5 and, therefore, $2y$ must be divisible by 5. As 2 is not divisible by 5, the second factor, y, has to be divisible by 5. This implies that y may only be equal to 5 (as already $y = 10$ yields $x = 0$ in equation (3.1), thus contradicting the assumption $x \neq 0$).

These are examples of word problems formulated within the boundaries of integer arithmetic. The problems can be solved without (grade-related) complexity of algebraic transformations through a recourse to an argument followed by a simple, perhaps purely mental, computation. That is how mathematical knowledge develops through a combination of argument and computation. Both argument and computation may be free from the extensive use of paper and pencil. Sometimes, argument can be supported by an image, so that "students can ... draw diagrams of important features and relationships" [*ibid*, p. 6]. The use of diagrams "can help support students in identifying the given quantities in a problem and the unknown quantity" [Ontario Ministry of Education, 2020, p. 122]. Furthermore, the use of diagrams enhances conceptual knowledge of students and makes them "more able to transfer this knowledge to new situations and apply it to new contexts"

[Department of Basic Education, 2018, p. 38]. The issue of knowledge transfer was discussed in the context of mathematical preparation of teachers in Singapore by acknowledging the challenge of finding the right balance between fostering students' success with standard tasks requiring correct application of procedural skills and developing their mastery of utilizing conceptual understanding when dealing with unfamiliar mathematical problems [Kaur, 2009].

In the above example of the book sale problem, the use of a linear equation in two unknowns was intentionally omitted. Linear equations, encountered at the upper elementary school level, are mathematical models of simple real-life situations when quantities of objects, depending on their properties, are put in two or more groups with either unknown cardinalities or unknown properties. As an example, consider the equation

$$5x + 2y = 15, \qquad (3.2)$$

in which x and y are unknowns to be found. Equation (3.2) can be contextualized in terms of the book sale already discussed above: *If Anna spent \$15 at a book sale buying \$5 hardcovers and \$2 softcovers, how many books of each type did Anna buy?* One can use a conceptual shortcut in solving equation (3.2) by noting that because both 15 and $5x$ (regardless of integer x) are divisible by 5, so is the term $2y$. This time, $2y$ may only be divisible by 5 if y is a multiple of 5. This implies $y = 5$ only (as already the value $y = 10$ makes $2y = 20 > 15$), whence $x = 1$. That is, Anna bought one hardcover book and five softcover books. Note that whereas replacing \$15 by \$25 would result in two solutions: five softcover and three hardcover books as well as ten softcover books and one hardcover book, replacing \$5 by \$4 would yield no solution (regardless, whether one spends \$15 or \$25).

However, a qualitatively different contextualization of equation (3.2) is possible: *If Anna spent \$15 at a book sale buying 2 hardcovers and 5 softcovers, what is the price of a book of each type?* This time, let us allow for the unknown prices to be both integers and non-integers. Whereas there is only one integer solution to equation (3.2), several non-integer solutions can also be found. For example, one such solution is $x =$ \$1.5 and $y =$ \$3.75. Another solution is $x =$ \$1.2 and $y =$ \$4.5. That is, when using a conceptual shortcut in solving an algebraic equation, one has to take into account contextual meaning of the unknowns involved.

In this regard, a geometric interpretation of a linear equation with two unknowns is worth mentioning. The graph of such an equation is a straight line in the (two-dimensional) coordinate plane passing through infinitely many points. Some straight lines such as the line described by equation (3.2) pass through a single point with positive integer coordinates and through infinitely many points with positive non-integer coordinates (Fig. 3.6). However, not all such points can serve as book prices. For example, the point (5/3, 10/3) belongs to the straight line; yet its coordinates, non-terminating periodic decimals $1.\overline{6}$ and $3.\overline{3}$, respectively, (see Chapter 8, Section 8.1), may not be used as book prices. Also, typically, a softcover is less expensive than a hardcover and therefore, we should assume $x < y$. Under such a condition, although the values $x = 2.4$ and $y = 1.5$ satisfy equation (3.2) and may represent book prices, in the context of the book sale of softcovers and hardcovers, such values have to be rejected as being not contextually coherent (see Chapter 6, Section 6.2).

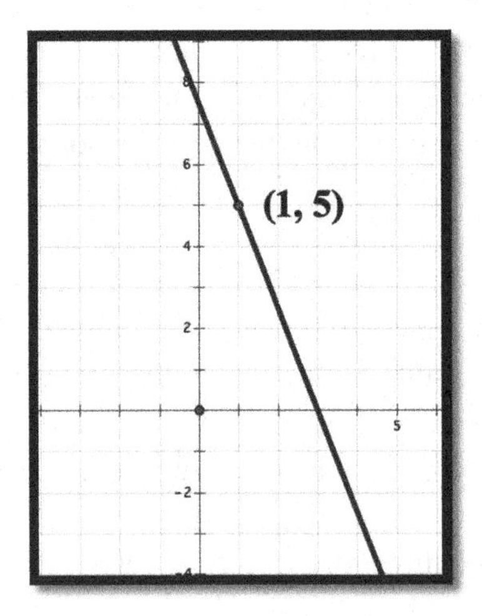

Fig. 3.6. The graph of equation (3.2) passes through the point (1, 5).

CHAPTER 4: DECOMPOSITION OF INTEGERS INTO LIKE ADDENDS

4.1 Decomposition in two addends

Additive decomposition of positive integers into like numbers can be put in context in order to begin with action and then describe this action in terms of arithmetic using the "**W**e **W**rite **W**hat **W**e See" (W^4S) principle [Abramovich, 2017]. For example, in order to build two towers out of ten square tiles (or linking cubes) and arrange them from the smallest to the greatest, one has to develop the set of towers (arranged in pairs) shown in Fig. 4.1. The W^4S principle yields the following numeric description of the elements of the set: $10 = 1 + 9 = 2 + 8 = 3 + 7 = 4 + 6 = 5 + 5$.

Sometimes, the order, in which two towers form a pair, matters and, thereby, different orders can be counted. (Alternatively, one can be asked to arrange towers both from the smallest to the greatest and from the greatest to the smallest). Mathematically, such an extension is due to the commutative property of addition enabling the above five decompositions to be augmented by the following four decompositions: $10 = 9 + 1 = 8 + 2 = 7 + 3 = 6 + 4$. Contextually, one can build nine pairs of towers out of 10 linking cubes. Mathematically, there are nine decompositions of the number 10 in two positive integer addends when counting their different orders.

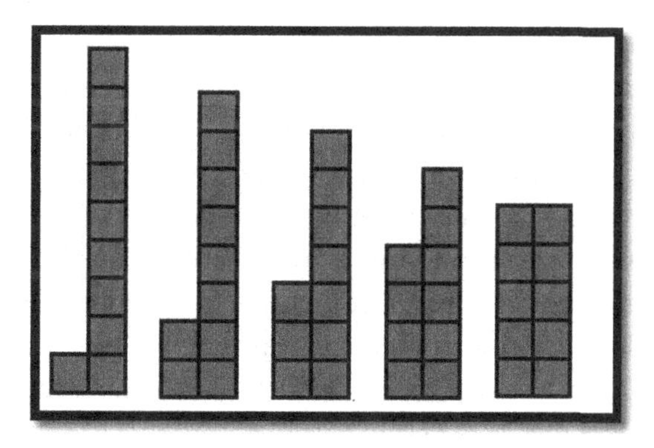

Fig. 4.1. Decomposition of 10 in two positive integer addends without regard to their order.

There is evidence, that young children, when asked to represent a number as a sum of two numbers, have difficulty in recognizing the existence of multiple representations. Indeed, as described in [Abramovich, 2021], a group of beginning second graders was asked to find all possible ways to represent the number 5 as a sum of two numbers, zero included, different order of numbers considered. It was expected that the students would provide the following response: $5 = 0 + 5 = 5 + 0 = 1 + 4 = 4 + 1 = 3 + 2 = 2 + 3$ (not necessarily in that order but, of course, using only positive integer decompositions as the second graders are not familiar with the *arithmetic* of other numbers; they are familiar with a few unit fractions though). However, as it turned out, the concept of the multiplicity of decompositions of integers into like addends was beyond understanding of the beginning second graders.

In order to rectify this challenge, the children were given five finger rings and asked to put them on their index and middle fingers in all possible ways and record their findings by completing drawings on pictures provided by the teacher. The outcome of this intervention was as follows. The students' pictures showed no ring on the index finger, five rings on the middle finger; five rings on the index finger, no ring on the middle finger; one ring on the index finger, four rings on the middle finger; one ring on the middle finger, four rings on the index finger; two rings on the index finger, three rings on the middle finger; two rings on the middle finger, three rings on the index finger. Furthermore, the students were able to describe their pictures through the above-mentioned representations of the number 5 using the W^4S principle.

Similarly, Max Wertheimer, one of the founders of Gestalt psychology, shared a case of a 9-year-old girl who was not successful in her studies of mathematics at school. In particular, she was unable to solve simple problems requiring the use of basic arithmetic. However, when given a problem which grew out of a concrete situation with which she was familiar and the solution of which "was required by the situation, she encountered no unusual difficulty, frequently showing excellent sense" [Wertheimer, 1959, pp. 273-274].

But what happened at the conclusion of the activities involving finger rings is worth noting. All of a sudden, a second-grade student asked: *How many ways can five rings be put on* ***three*** *fingers*? That is, young children, when learning mathematics in context familiar to them, are capable not only to be successful with the subject matter, but better still, they can naturally ask quite simple questions about this context the answers to which may not be available to teaches right away (if at all). A grade-appropriate answer to the above question will be given at the end of the next section.

4.2 Decomposition in three addends

The above story about the second-grade classroom can serve as a motivation to consider additive decomposition of positive integers in three like numbers. To this end, a strategy similar to the one used in the context of rings and fingers, which may not be obvious without having a certain problem-solving experience, can be used to build three towers out of ten square tiles (or linking cubes) and arrange them from the smallest to the greatest[10]. Such towers are shown in Fig. 4.2 and the strategy used here is to begin with the highest possible tower and then gradually decrease its size by adding each new tile to a smaller tower keeping in place the condition of their arrangement from the smallest to the greatest. Now, the diagram of Fig. 4.2 enables one to see how to decompose 10 into three addends arranged from the least to the greatest: $10 = 1 + 1 + 8 = 1 + 2 + 7 = 1 + 3 + 6 = 2 + 2 + 6 = 1 + 4 + 5 = 2 + 3 + 5 = 2 + 4 + 4 = 3 + 3 + 4$.

Note that this strategy does not provide the number of ways to arrange the towers in all possible orders. However, Fig. 4.2 can be used to explore how many new towers stem from each of the eight towers. To this end, one has to recognize an important difference in the towers. Namely, either a set of three towers has them all of different height or two of the towers are of the same height. (Towers may not be all of the same height as 10 is not a multiple of 3). In the latter case, the tallest tower can be positioned among two identical towers in three different ways. In the former case, counting is more complicated. As an example, consider the second triple of towers in Fig. 4.2. If the tallest tower is located at the far

[10] In terms of rings and fingers, the activity is equivalent to putting ten finger rings on three fingers in such a way that for any pair of fingers, a finger on the left may not have more rings than the finger on the right and each finger has a ring. The second grader, apparently, asked a question about a slightly more complicated distribution of rings when, in particular, all the rings may be put on any of the three fingers (see Remark 4.1).

right, then the other two towers can be positioned in two ways. If the tallest tower is positioned at the far left, then the remaining two towers can also be positioned in two ways. Finally, the tallest tower can be positioned between the other two towers. In that case, these smaller towers can also be positioned in two ways. This reasoning is explained in Fig. 4.3 for the triple (1, 2, 7). Therefore, according to Fig. 4.2, we have four triples of the latter type – (1, 1, 8), (2, 2, 6), (2, 4, 4), (3, 3, 4) – and four triples of the former type – (1, 2, 7), (1, 3, 6), (1, 4, 5), (2, 3, 5). From each of the four triples of the latter type three towers result, giving the total of 12 towers. From each of the four triples of the former type six towers result, giving the total of 24 towers. In all, we have 36 towers. That is, the number 10 can be decomposed into the sums of three positive integers in all possible orders in 36 ways. In other words, by physically permuting towers in each of the eight triples of towers listed from left to right, the number of permutations is represented, respectively, by the sequence 3, 6, 6, 3, 6, 6, 3, 3, the sum of the terms of which is 36. An alternative counting technique allowing one to obtain this sequence of numbers will be discussed in Chapter 11, Section 11.4.

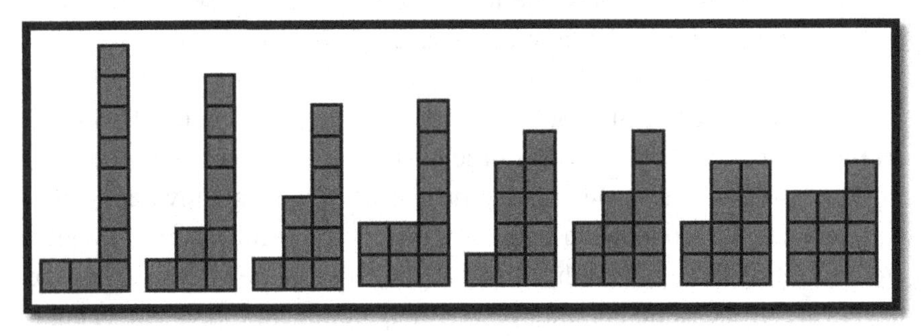

Fig. 4.2. Additive decomposition of 10 in three parts (focus on the largest tower).

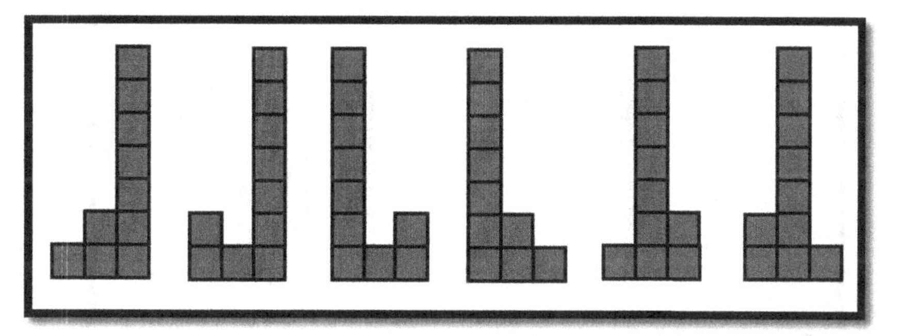

Fig. 4.3. Counting permutations of three different towers.

Remark 4.1. An answer to the second-grade student's question about the number of ways five finger rings can be put on three fingers is shown in Fig. 4.4 where three different situations are considered. In the first situation, only one finger does not have a ring implying that other two fingers may have either one and four rings or two and three rings. In both cases, all three fingers have different number of rings and therefore each case yields six different ways to put five rings on three fingers. In the second situation, putting a single ring on a finger allows for other two fingers to have either one and three rings or two and two rings. Each case yields three different ways to put five rings on three fingers. Finally, the third situation is when all five rings are put on one finger leaving the other two fingers without a ring. This yields three new ways to put five rings on three fingers. Therefore, there are 21 ways to put five rings on three fingers. In Chapter 11, Section 11.5, the number 21 will be found using a different strategy.

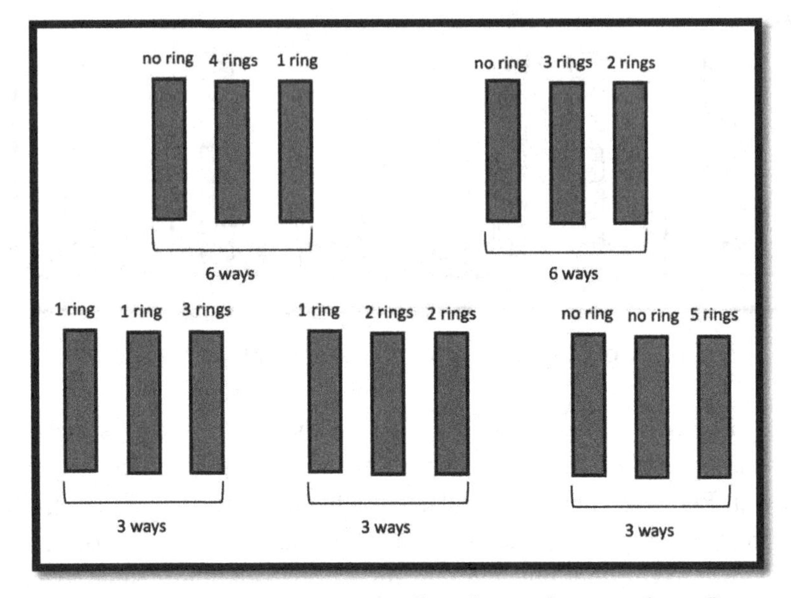

Fig. 4.4. There are 21 ways to put five finger rings on three fingers.

4.3 Alternative decomposition strategies

The third strategy to decompose 10 in three (positive integer) addends (under the previous condition of arranging towers in a non-decreasing order) stems from the diagram of Fig. 4.5. This time, instead of focusing on the tallest tower in a triple (like in Fig. 4.2), one can focus on the base of a triple. Fig. 4.5 shows three different bases: a base with three blocks (four triples of towers), a base with six blocks (three triples of towers), and a base with nine blocks (one triple of towers). In the first case, the problem of decomposition of 10 in three addends is reduced to that of decomposition of seven ($10 - 3 = 7$) in fewer than three addends ($7 = 1 + 6 = 2 + 5 = 3 + 4$). In the second case, the problem of decomposition of 10 in three addends is reduced to that of decomposition of four ($10 - 6 = 4$) in fewer than three addends ($4 = 1 + 3 = 2 + 2$). In the third case of the nine-block base only one block ($10 - 9 = 1$) remains.

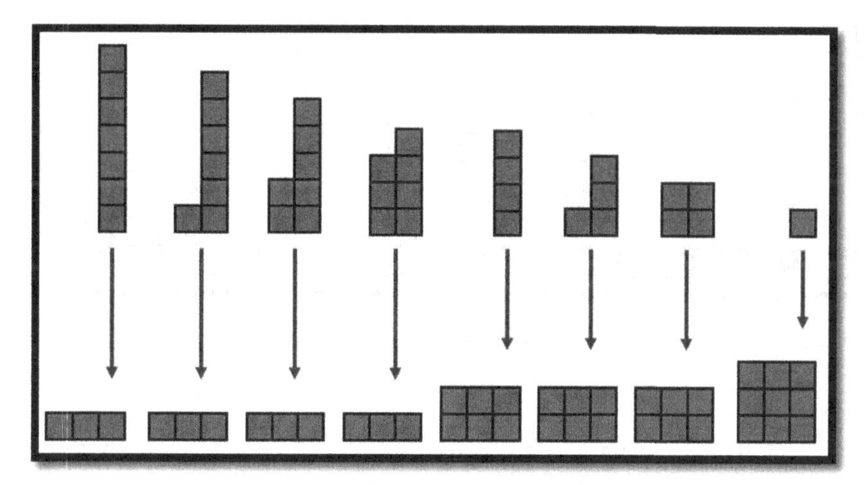

Fig. 4.5. Additive decomposition of 10 in three parts (focus on the triple's base).

The fourth strategy of decomposing 10 in three (positive integer) addends using the context of towers is to focus on the smallest tower and decompose the remaining blocks in two addends similar to how it was done in the case of Fig. 4.1. Thus, we have (Fig. 4.6) four sets of three towers with one block used for the first tower: $10 = 1 + 1 + 8 = 1 + 2 + 7 = 1 + 3 + 6 = 1 + 4 + 5$; three sets of three towers with two blocks used as the first tower: $10 = 2 + 2 + 6 = 2 + 3 + 5 = 2 + 4 + 4$; and a single set with three blocks used as the first tower: $10 = 3 + 3 + 4$. An emphasis on multiple strategies in decomposing an integer into three summands is aimed at the metacognitive development of students, something that includes their ability to *select* and *use* an appropriate (in their view) problem-solving strategy when dealing with a non-routine problem [Ministry of Education Singapore, 2020].

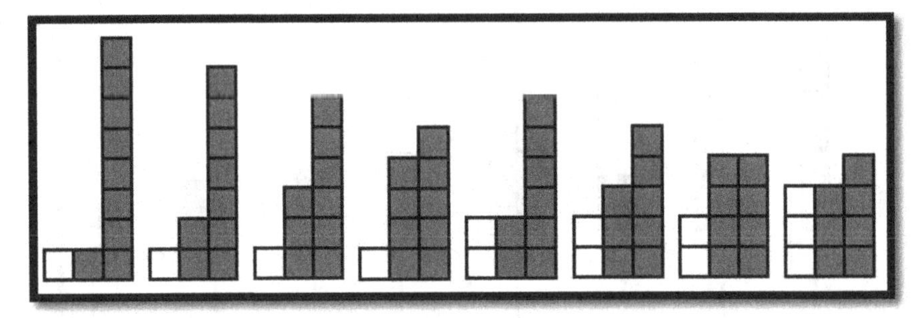

Fig. 4.6. Additive decomposition of 10 in three parts (focus on the smallest tower).

Fig. 4.7, using the first strategy of decomposing 10 in three addends, shows nine decompositions of 10 into *four* addends arranged from the least to the greatest focusing on the largest (the fourth) tower: $10 = 1 + 1 + 1 + 7 = 1 + 1 + 2 + 6 = 1 + 1 + 3 + 5 = 1 + 2 + 2 + 5 = 1 + 1 + 4 + 4 = 1 + 2 + 3 + 4 = 2 + 2 + 2 + 4 = 1 + 3 + 3 + 3 = 2 + 2 + 3 + 3$.

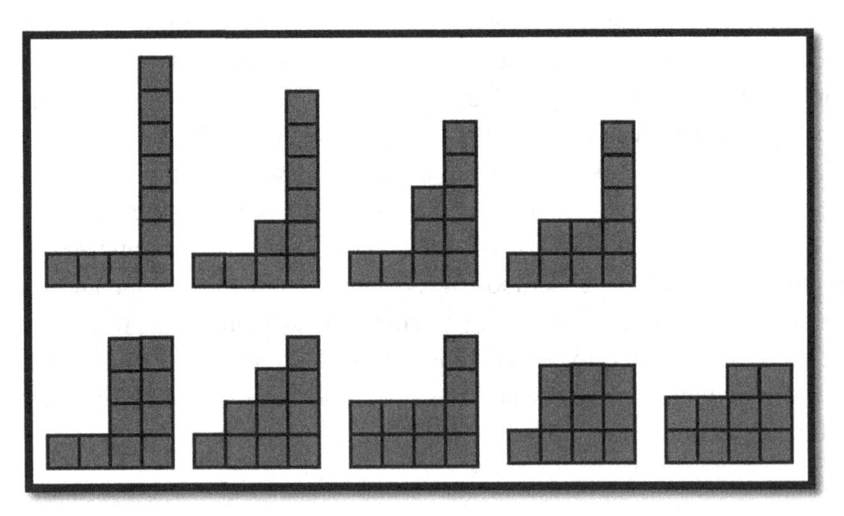

Fig. 4.7. Additive decomposition of 10 in four parts (focus on the largest tower).

Another activity may deal with the pattern shown in Fig. 1.14 (Chapter 1, Section 1.4). Whereas the sum of blocks in a pattern of a certain length is unknown, one can talk about the properties of this sum, based on the rule through which the pattern develops. First, in the context of Fig. 1.14, whatever the sum, it can always be additively decomposed into the multiples of six. Second, each triple of towers can be written as (k, $2k$, $3k$) and whatever k, the triple can be turned into the pair ($3k$, $3k$). Because each element of the pair is a multiple of 3, the sum is divisible by 3. Furthermore, because each element of the pair is the same, the sum is divisible by 2. Therefore, the sum is divisible by 6. This can be demonstrated physically through the rearrangement of blocks into the pairs and triples of equal size towers. In order to find the number of blocks in the 100th tower in the pattern shown in Fig. 1.14, one can note that the largest value of k for which $3k < 100$ is $k = 33$; therefore the 34th triple of towers begins with the 100th tower being 34-block tall.

4.4 Decomposition through creating spaces among addends

Consider the triples of towers shown in Fig. 4.2. Let us represent each tower in the form of a train; that is, placing the blocks of each tower horizontally (Fig. 4.8). In doing so, each element of a triple has to be separated from another element in order to have triples of trains. One can see that there are two spaces that separate any three trains. This may prompt new idea of decomposition of a number in three addends: find all ways to create two spaces among the blocks representing the number to be additively decomposed.

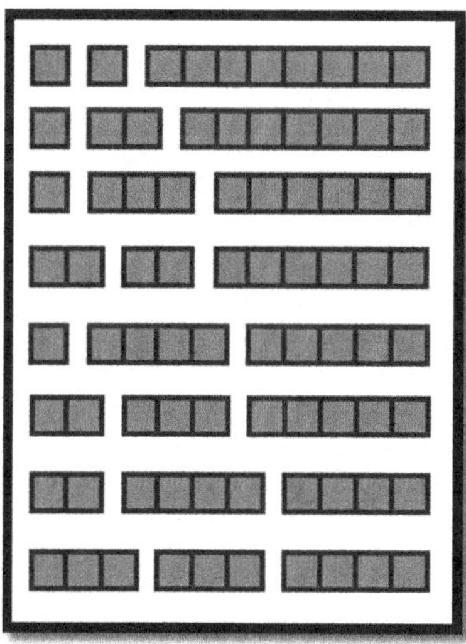

Fig. 4.8. Alternative decomposition of 10 in three addends.

Whereas the trains in Fig. 4.8 are arranged from left to right in the non-decreasing lengths of the trains, finding all ways to create two spaces among ten blocks would include all different orders in which three trains are positioned from left to right. Similar to Fig. 4.2, in Fig. 4.8 we have four triples of trains, two of which have equal lengths, and four triples of trains of different lengths. By analogy, when decomposing ten in four (positive integer) addends in all possible orders, one has to find all ways to create three spaces among ten blocks in order to form four trains. In Chapter 11, this idea will be further elaborated to include the concept of permutations of letters in a word.

CHAPTER 5: ACTIVITIES WITH ADDITION AND MULTIPLICATION TABLES

5.1 Geometrization of arithmetical tables

The multiplication table is well known as part of elementary school mathematics curriculum. Its traditional educational uses are purely arithmetical, stemming from the need to succinctly record repeated addition in the form of the corresponding multiplication facts. The multiplication table can also be used in solving simple division problems when one out of two factors forming a known product is unknown. A few conceptually oriented activities with the multiplication table were discussed in Chapter 2, Section 2.3. In visual terms, multiplication (alternatively, repeated addition) can be represented through a concrete activity when integer-sided rectangles have to be constructed by using unit squares. In that case, the product of two integers can be understood as the number of unit squares inside a rectangle so constructed, and therefore, conceptually, this product can be interpreted as the area of the rectangle expressed in terms of the number of unit squares used to measure space inside the rectangle (see Chapter 3, Fig. 3.2). Another geometric characteristic associated with a rectangle is perimeter. In the context of a rectangle filled with unit squares, its perimeter can be defined as the number of outer edges of the unit squares forming the border of the rectangle.

Similar to the case of area, one can consider only a pair of adjacent sides as the essential characteristics of a rectangle; the number of edges of unit squares forming such a pair of sides is one-half of the perimeter (alternatively, semi-perimeter). The sums of two integers can form an addition table. Therefore, using the concepts of area and semi-perimeter, both the multiplication and the addition tables can be explored within a unified context of arithmetic and geometry.

5.2 Counting all numbers in addition and multiplication tables

Arithmetical tables, in general, represent a conceptually rich milieu of activities for fostering mathematical reasoning skills using a numerical approach and serving as a springboard into the development of algebraic thinking, one of the domains included in the Common Core State Standards [2010, p. 10] as early as at the kindergarten level through a symbolic description of patterns to "express regularity in repeated reasoning". The multiplication and addition tables are examples of such a milieu available to students at the post-kindergarten level. One activity with the tables deals with answering the following (summation-oriented) questions: *What is the sum of all numbers in the* 10×10 *multiplication table? What is the sum of all numbers in the* 10×10 *addition table?*

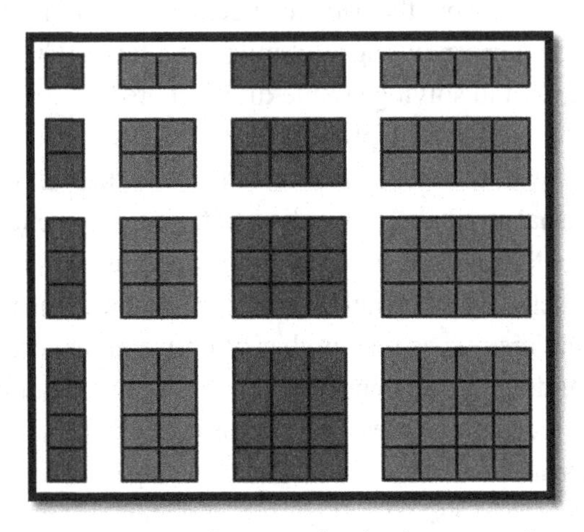

Fig. 5.1. There are 16 types of rectangles in the range from 1×1 to 4×4.

Before answering the above questions, let us create two tables with all types of products and sums ranging in size from 1×1 to 4×4 and find areas and semi-perimeters for all the rectangles. In that case, we have the products 1×1, 1×2, 1×3, 1×4, 2×1, 2×2, 2×3, 2×4, 3×1, 3×2, 3×3, 3×4, 4×1, 4×2, 4×3, 4×4 and the sums $1 + 1$, $1 + 2$, $1 + 3$, $1 + 4$, $2 + 1$, $2 + 2$, $2 + 3$, $2 + 4$, $3 + 1$, $3 + 2$, $3 + 3$, $3 + 4$, $4 + 1$, $4 + 2$, $4 + 3$, $4 + 4$, forming, respectively, the multiplication and the addition tables of size 4×4. A recourse to such small size tables makes it possible to develop summation strategies that can be applied to larger tables. Fig. 5.1 shows all 16 types of the rectangles related to the above products and sums. The corresponding tables, generated by *Wolfram Alpha*, are shown in Figs 5.2 and 5.3, respectively.

×	1	2	3	4
1	1	2	3	4
2	2	4	6	8
3	3	6	9	12
4	4	8	12	16

Fig. 5.2. The 4×4 multiplication table generated by *Wolfram Alpha*.

+	1	2	3	4
1	2	3	4	5
2	3	4	5	6
3	4	5	6	7
4	5	6	7	8

Fig. 5.3. The 4×4 addition table generated by *Wolfram Alpha*.

How can one find the sum of all numbers in the 4×4 multiplication table shown in Fig. 5.2? One way to find the sum is pretty straightforward: find the sum of numbers in each of the four rows (or columns). We have $1 + 2 + 3 + 4 = 10$, $2 + 4 + 6 + 8 = 20$, $3 + 6 + 9 + 12$

$= 30$, $4 + 8 + 12 + 16 = 40$, so that the sum of all numbers is equal to $10 + 20 + 30 + 40 = 100$.

The above way of finding the sum of all numbers in the 4×4 multiplication table can prompt the idea that the sum of numbers in each row (column) is a multiple of the sum in the first row (column) where the corresponding multiplier is equal to the row's (column's) number. That is, the second way of finding the sum of all numbers in the 4×4 multiplication table can be as follows:

$$(1 + 2 + 3 + 4) + (2 + 4 + 6 + 8) + (3 + 6 + 9 + 12) + (4 + 8 + 12 + 16)$$
$$= 1 \times (1 + 2 + 3 + 4) + 2 \times (1 + 2 + 3 + 4) + 3 \times (1 + 2 + 3 + 4)$$
$$+ 4 \times (1 + 2 + 3 + 4) = (1 + 2 + 3 + 4) \times (1 + 2 + 3 + 4)$$
$$= (1 + 2 + 3 + 4)^2 = 10 \times 10 = 100.$$

The third way of finding the sum stems from an observation that in each row (column) of the table the sums of numbers equidistant from its borders are the multiples of five, and there are two such sums in each row (column). Therefore, the sum of all numbers in the four rows (columns) of the 4×4 multiplication table can be found as follows:

$$2 \times 5 + 2 \times 10 + 2 \times 15 + 2 \times 20 = 2 \times 5 \times (1 + 2 + 3 + 4) = 10 \times 10 = 100.$$

The fourth way of finding the sum is to take advantage of the commutative property of multiplication due to which the numbers above and below the main (top-left/bottom-right) diagonal of the table are the same. That is, the sum of all numbers in the 4×4 multiplication table is equal to $(1 + 4 + 9 + 16) + 2[(2 + 6 + 12) + (3 + 8) + 4] = 30 + 70 = 100$.

Finally, the fifth way of finding the sum is less obvious than the previous four ways. It requires some kind of insight. To this end, one can do summation within the table's gnomons (see Chapter 1, Section 1.2) as follows: $1 + (2 + 4 + 2) + (3 + 6 + 9 + 6 + 3) + (4 + 8 + 12 + 16 + 12 + 8 + 4) = 1 + 8 + 27 + 64 = 100$. Note that $1 = 1^3$, $8 = 2^3$, $27 = 3^3$, $64 = 4^3$, and, therefore, by connecting the fifth and the second ways of summation, the relationship $1^3 + 2^3 + 3^3 + 4^3 = (1 + 2 + 3 + 4)^2$ between the square of the sum of the first four positive integers and their cubes can be derived. One can see that this insight is not just another way of solving a problem. It allowed for an unexpected relationship to be discovered in a special case prompting its extension to other similar cases and, eventually, leading to generalization. Once again, emphasizing multiple ways of solving a non-routine problem helps learners of mathematics to become metacognitive

[Ministry of Education Singapore, 2020] as one's ability to think about thinking promotes reflection, which, in turn, "impels to inquiry" [Dewey, 1933, p. 7].

Remark 5.1. In order to find the sum of all numbers in the 5×5 multiplication table, one can use a conceptual shortcut by noting that transition to this table from the 4×4 multiplication table requires adding a gnomon of rank five the sum of numbers in which is $5^3 = 125$. Therefore, the sum of all numbers in the 5×5 multiplication table is equal to $100 + 125 = 225$. This number can be confirmed by finding that $(1 + 2 + 3 + 4 + 5)^2 = 15^2 = 225$. Alternatively, one can write

$$1^3 + 2^3 + 3^3 + 4^3 + 5^3 = (1 + 2 + 3 + 4 + 5)^2.$$

The above five summation strategies can be used in the case of the 10×10 multiplication table. The first way leads to the following ten sums (see Chapter 3, Section 3.3) $1 + 2 + ... + 10 = 55$, $2 \times (1 + 2 + ... + 10) = 2 \times 55$, $3 \times (1 + 2 + ... + 10) = 3 \times 55$, ..., $10 \times (1 + 2 + ... + 10) = 10 \times 55$, which, when added together, yield $55 \times (1 + 2 + 3 + ... + 10) = 55^2 = 3025$. Fig. 5.4 shows how to square 55 using a conceptual shortcut (see Chapter 3, Section 3.2, Figs 3.1 and 3.2).

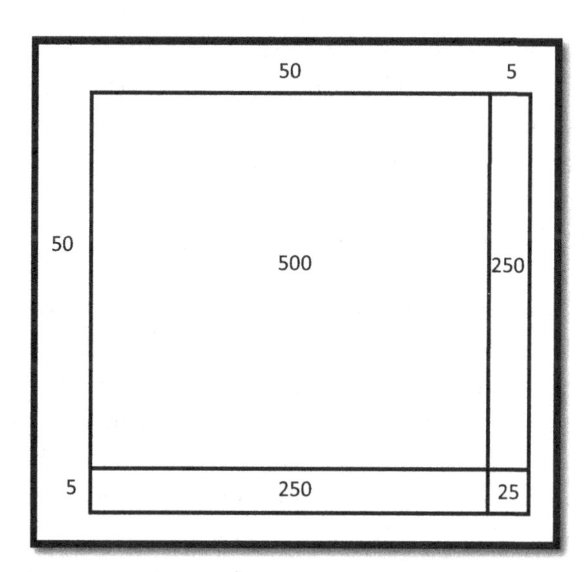

Fig. 5.4. Finding 55^2 through a conceptual shortcut.

In order to find the sum of all numbers in the 4×4 addition table, five different strategies can be used as well. The first strategy is to note that in each row of the table (beginning from the second row) the sum of the numbers is greater than the sum in the previous row by four. Therefore, one can find the sum of all numbers in the table as follows:

$$4 \times (2 + 3 + 4 + 5) + 4 + 8 + 12 = 80.$$

The second strategy is to add numbers located along the bottom-left/top-right diagonals. That is, the sum is equal to $(5 + 5 + 5 + 5) + [2 + (3 + 3) + (4 + 4 + 4) + (6 + 6 + 6) + (7 + 7) + 8] = 80$.

The third strategy is to add numbers located along the top-left/bottom-right diagonals. This time, the sum can be calculated as follows:

$$(2 + 4 + 6 + 8) + 2 \times [(3 + 5 + 7) + (4 + 6) + 5] = 80.$$

The fourth strategy is to note that the sums of two numbers equidistant from the borders of the table in each row are, respectively, 7, 9, 11, and 13. Therefore, the sum of all numbers in the table is equal to

$$2 \times (7 + 9 + 11 + 13) = 2 \times 40 = 80.$$

The fifth strategy is to add the numbers in the table's gnomons:

$$2 + (3 + 4 + 3) + (4 + 5 + 6 + 5 + 4) + (5 + 6 + 7 + 8 + 7 + 6 + 5) = 80.$$

Learning to solve a problem in more than one way is encouraged by several modern-day educational documents associated with mathematical preparation of prospective elementary school teachers around the world and it allows for more opportunities to connect mathematics across different grades [Department of Basic Education, 2018; Association of Mathematics Teacher Educators, 2017; Conference Board of the Mathematical Sciences, 2012; Ministry of Education Singapore, 2020; Ontario Ministry of Education, 2020; Hwang and Han, 2013; Takahashi et al., 2004]. Connecting mathematics across the grades through problem-solving activities highlights the dual nature of the activities as an instrument of pragmatic and epistemic development of learners. The pragmatic nature of problem solving operates in the direction of a problem to be solved. The epistemic nature of problem solving operates in the direction of a problem solver by supporting their "collateral learning" [Dewey, 1938, p. 49] opportunities.

Remark 5.2. One may note (Fig. 5.5) that the sum of numbers located below the bottom-left/top-right diagonal is twice the sum of numbers located above it. Indeed,

$$(12 + 22 + 30 + 36 + 40) = 2 \cdot (2 + 6 + 12 + 20 + 30).$$

This pattern continues as we consider larger square size addition tables.

Remark 5.3. Numbers located above and below the bottom-left/top-right diagonal of a square size addition table form a triangle the entries of which develop through an interesting pattern. Consider the 6 × 6 addition table (Fig. 5.5). Forming a triangle evolving along the diagonals parallel to the main (top-left/bottom-right) diagonal of the table results in the triangle pictured in Fig. 5.6. The triangle resembles the famous Pascal's triangle[11] (Chapter 12, Section 12.3, Fig. 12.5). The non-lateral entries of the triangle are half the sums of the entries immediately above them.

+	1	2	3	4	5	6
1	2	3	4	5	6	7
2	3	4	5	6	7	8
3	4	5	6	7	8	9
4	5	6	7	8	9	10
5	6	7	8	9	10	11
6	7	8	9	10	11	12

Fig. 5.5. A 6 × 6 addition table generated by *Wolfram Alpha*.

[11] Blaise Pascal (1623-1662), a French mathematician, physicist, and philosopher, one of the founders of probability theory.

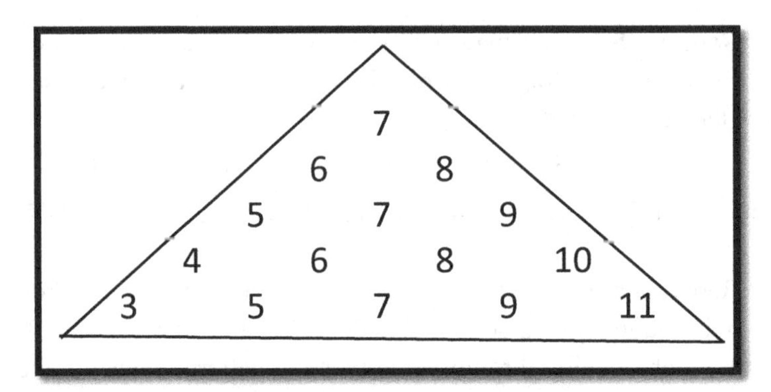

Fig. 5.6. A Pascal-like triangle formed by the entries of the 6 × 6 addition table.

Remark 5.4. The fifth strategy of adding numbers within the gnomons of an addition table is worth exploring in more depth. Consider the addition table of Fig. 5.5. The sum of numbers in the gnomon of rank four, after noting that 8 = 4 + 4, is twice the sum 4 + 5 + 6 + 7 = 22. Fig. 5.7 shows that this sum can be represented by another arithmetic sequence which starts with 1 and has difference 3. That is, 4 + 5 + 6 + 7 = 1 + 4 + 7 + 10. Likewise, after noting that 10 = 5 + 5, the sum of numbers in the gnomon of rank five is twice the sum 5 + 6 + 7 + 8 + 9 = 35, which, as shown in Fig. 5.8, can be represented as the sum of the first five terms of the same arithmetic sequence through which the numbers in the gnomon of rank four were represented; that is, 5 + 6 + 7 + 8 + 9 = 1 + 4 + 7 + 10 + 13. One can check to see that in the gnomons of ranks two and three the following equalities can be developed: 2 + 3 = 1 + 4 = 5 and 3 + 4 + 5 = 1 + 4 + 7 = 12. Thus, the sums of numbers in the gnomons of an addition table are partial sums of the arithmetic sequence 1, 4, 7, 10, 13, ...These partial sums are known as pentagonal numbers. Recall, that the triangular and square numbers are, respectively, the partial sums of the sequences 1, 2, 3, 4, 5, ... and 1, 3, 5, 7, 9, ... In much the same way, the pentagonal numbers 1, 5, 12, 22, 35, ... (halves of the sums of numbers in the gnomons of an addition table) are the partial sums of the sequence 1, 4, 7, 10, 13, ... Fig. 5.9 shows how the last sequence of numbers makes pentagons. This unexpected connection of the addition table to pentagonal numbers demonstrates that, in the words of Australian educators, “mathematics has its own value and beauty and it is intended that students will appreciate the elegance and power of mathematical thinking, experience mathematics as enjoyable, and encounter teachers who communicate this enjoyment" [National Curriculum Board, 2008, p. 2].

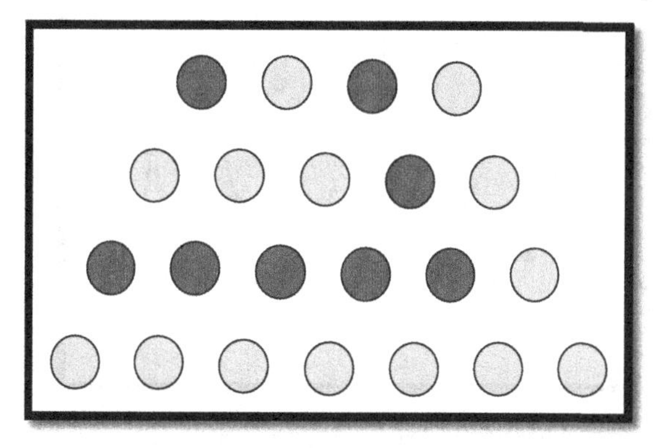

Fig. 5.7. Image of the equality $4 + 5 + 6 + 7 = 1 + 4 + 7 + 10$.

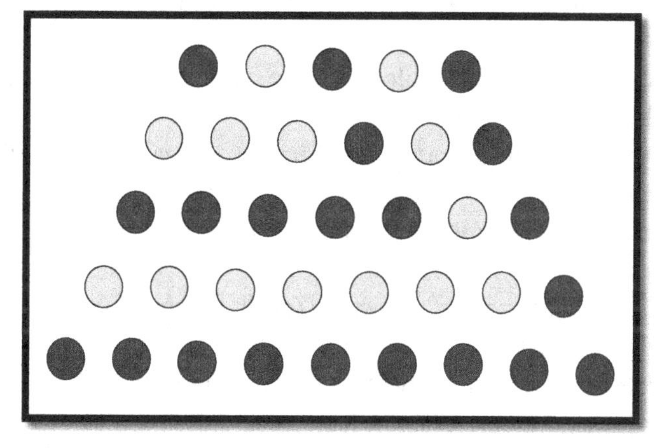

Fig. 5.8. Image of the equality $5 + 6 + 7 + 8 + 9 = 1 + 4 + 7 + 10 + 13$.

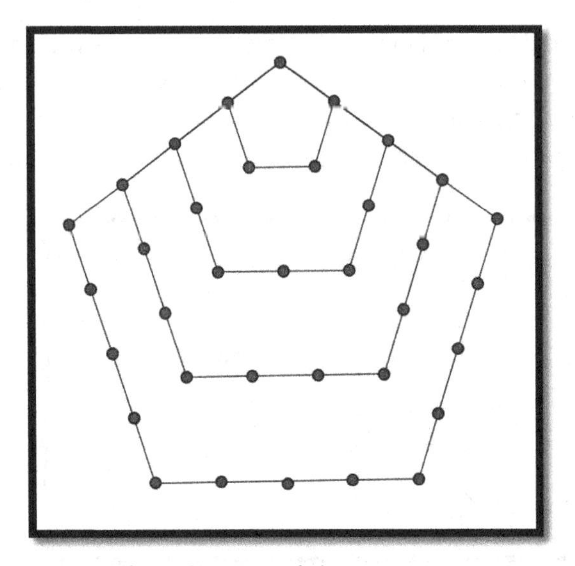

Fig. 5.9. An evolving pentagon.

5.3 Counting numbers with special properties in multiplication and addition tables

Other questions that can be explored in the context of multiplication and addition tables are as follows:

How many even (odd) numbers are there in the 10×10 *multiplication table?*

How many even (odd) numbers are there in the 10×10 *addition table?*

What is the sum of all even (odd) numbers in the 10×10 *multiplication table?*

What is the sum of all even (odd) numbers in the 10×10 *addition table?*

To begin answering these questions, a smaller size tables can be considered first. *How many even numbers are there in the* 4×4 *multiplication table?* In an even row (column) of the table all numbers are even; in an odd row of the table, only half of numbers are even (Fig. 5.2). Therefore, out of 16 numbers in the table, 12 numbers are even, and 4 numbers are odd. Likewise, out of 100 numbers in the 10×10 multiplication table, each row of an even rank has 10 even numbers; each row of an odd rank has 5 even numbers. Therefore, in the 10 × 10 multiplication table there are 75 even numbers and 25 odd numbers.

In order to find the sum of even numbers in the 4 × 4 multiplication table, one can first find the sum of odd numbers in the table, keeping in mind that the sum of all numbers is 100 (Section 5.2). The sum of odd numbers is equal to $(1 + 3) + 3(1 + 3) = (1 + 3)(1 + 3) = 16$. Therefore, the sum of even numbers in the table is equal to $100 - 16 = 84$. A similar strategy can be used to find the sum of even numbers in the 10 × 10 multiplication table. The spreadsheet of Fig. 5.10 (the programming of which is included in Chapter 13) shows the table displaying odd products only. Their sum can be found as follows:

$1 \times (1 + 3 + 5 + 7 + 9) + 3 \times (1 + 3 + 5 + 7 + 9) + 5 \times (1 + 3 + 5 + 7 + 9) + 7 \times (1 + 3 + 5 + 7 + 9) + 9 \times (1 + 3 + 5 + 7 + 9) = 25 \times (1 + 3 + 5 + 7 + 9) = 25 \times 25 = 625$

Then, the sum of even numbers in the 10 × 10 *multiplication* table is equal to $3025 - 625 = 2400$.

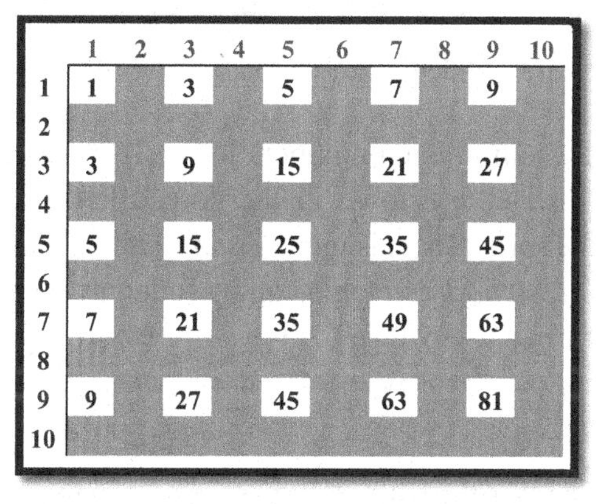

	1	2	3	4	5	6	7	8	9	10
1	1		3		5		7		9	
2										
3	3		9		15		21		27	
4										
5	5		15		25		35		45	
6										
7	7		21		35		49		63	
8										
9	9		27		45		63		81	
10										

Fig. 5.10. Odd products in the 10 × 10 multiplication table.

The sum of even numbers in the 10 × 10 *addition* table (Fig. 5.11) can be found as follows. Each row of the table has 5 even numbers (shown in Fig. 5.11) and 5 odd numbers (not shown (shaded) in Fig. 5.11). Therefore, there are 50 even numbers and 50 odd numbers in the 10 × 10 addition table. The spreadsheet of Fig. 5.11 (the programming of which is

included in Chapter 13) shows such a table with even numbers displayed. The sum of all such numbers can be found by adding numbers along the bottom-left/top-right diagonals. Using the diagram of Fig. 5.11, calculations can be carried out as follows:
$2 + 3 \times 4 + 5 \times 6 + 7 \times 8 + 9 \times 10 + 9 \times 12 + 7 \times 14 + 5 \times 16 + 3 \times 18 + 20 = 1 \times (2 + 20) + 3 \times (4 + 18) + 5 \times (6 + 16) + 7 \times (8 + 14) + 9 \times (10 + 12) = 22 \times (1 + 3 + 5 + 7 + 9) = 22 \times 25 = 550$.

	1	2	3	4	5	6	7	8	9	10
1	2		4		6		8		10	
2		4		6		8		10		12
3	4		6		8		10		12	
4		6		8		10		12		14
5	6		8		10		12		14	
6		8		10		12		14		16
7	8		10		12		14		16	
8		10		12		14		16		18
9	10		12		14		16		18	
10		12		14		16		18		20

Fig. 5.11. Even sums in the 10×10 addition table.

That is, the sum of even numbers in the 10×10 *addition* table is equal to 550. Likewise, as shown in the spreadsheet of Fig. 5.12 (the programming of the spreadsheet is included in Chapter 13), the sum of odd numbers in the 10×10 *addition* table can be found as follows:
$2 \times 3 + 4 \times 5 + 6 \times 7 + 8 \times 9 + 10 \times 11 + 8 \times 13 + 6 \times 15 + 4 \times 17 + 2 \times 19 = 2 \times (3 + 19) + 4 \times (5 + 17) + 6 \times (7 + 15) + 8 \times (9 + 13) + 10 \times 11 = 22 \times 2 \times (1 + 2 + 3 + 4) + 22 \times 5 = 22 \times 25 = 550$.

One can see that the sum of even numbers in the 10×10 addition table is the same as the sum of odd numbers in that table. One may wonder: what is a conceptual explanation of such a phenomenon? In order to provide a conceptual explanation of a phenomenon discovered computationally within a numeric table, one has to think about the very genesis of the emergence of such a table. The genesis was in geometry – addition tables were constructed by recording semi-perimeters of rectangles built from unit squares. First note that within any quadruple of entries of the table, there are two even numbers and two odd numbers with the same sum. For example, the top-left corner of the table pictured in Fig. 5.5 has the quadruple 2, 3, 3, 4 where $2 + 4 = 3 + 3$. Same relationship can be observed within other quadruples of the entries of the table. Second, this observation can be explained in general terms using the diagram of Fig. 5.13 which includes four rectangles, the semi-perimeters of which are the entries of an arbitrary quadruple of entries in any addition table. One can see that the rectangles have the pair of side lengths (a, b), $(a + 1, b)$, $(a, b + 1)$ and $(a + 1, b + 1)$ with semi-perimeters $a + b$, $a + b + 1$, $a + b + 1$ and $a + b + 2$. Here, the numbers $a + b$ and $a + b + 2$ are of the same parity as they differ by the number 2. At the same time, if $a + b$ is even, then $a + b + 1$ is odd and vice versa. Furthermore, $(a + b) + (a + b + 2) = (a + b + 1) + (a + b + 1)$. Finally, it should be noted that the equality between the sums of even and odd entries of an addition table is only true for square size tables of an even size. Indeed, already in the gnomon of rank 3 of the addition table pictured in Fig. 5.3, the sum of even entries is 14 and the sum of odd entries is 10. So, the transition from the 2×2 addition table to the 3×3 addition table does not preserve the equality between the sums of even and odd entries observed in the former table.

	1	2	3	4	5	6	7	8	9	10
1		3		5		7		9		11
2	3		5		7		9		11	
3		5		7		9		11		13
4	5		7		9		11		13	
5		7		9		11		13		15
6	7		9		11		13		15	
7		9		11		13		15		17
8	9		11		13		15		17	
9		11		13		15		17		19
10	11		13		15		17		19	

Fig. 5.12. Odd sums in the 10×10 addition table.

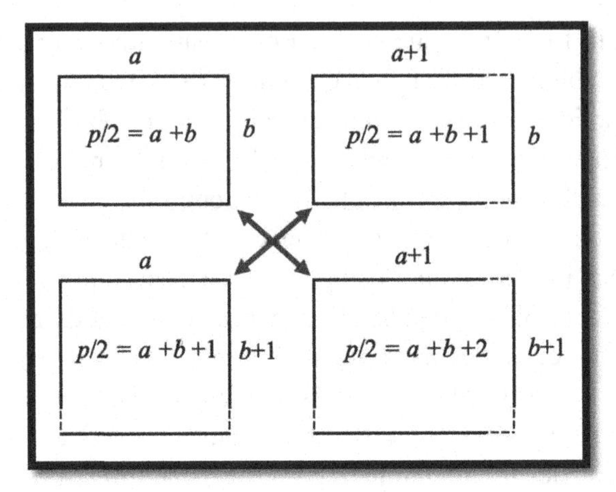

Fig. 5.13. Geometric interpretation of a quadruple of entries in an addition table.

CHAPTER 6: USING TECHNOLOGY IN POSING AND SOLVING PROBLEMS

6.1 Technology as a link between problem solving and problem posing

The use of technology in mathematical problem posing can be characterized as a cultural support of designing curriculum materials for a mathematics classroom. Such a design may include reformulation of existing problems or formulation of problems that are new for the traditional curriculum. Through this creative process, a teacher can use various tools of technology developed by advanced members of the modern-day technological society for a wide range of activities. For example, an electronic spreadsheet, originally developed for non-educational purposes [Power, 2000], has been used effectively in mathematics education [e.g., Baker and Sugden, 2003; Neuwirth and Arganbright, 2004; Abramovich, 2016]. In particular, in the specific context of mathematical problem posing and solving, one can put to work the power of this computational tool in order to generate a solution to a self-posed grade-appropriate problem and to see whether the solution is unique, or whether it exists at all[12]. That is, teacher's knowledge of mathematics in the age of technology should include at least basic skills in programming a spreadsheet to enable a numeric problem to be both posed and solved. This implies that computing technology, in general, and an electronic spreadsheet, in particular, make problem posing and problem solving being inherently linked to each other. The important task for a teacher is to appreciate this relation between problem posing and problem solving, to learn how to recognize the emergence of a new problem through a teacher-student interaction, and to "be expert in ... the craft of task design" [Conference Board of the Mathematical Sciences, 2012, p. 65]. The use of the word expert in the last quote is supported by the viewpoint that when dealing with "the real world, most of the time, an answer is easier than defining a question" [Dyson, 2012, p. 163]. To have first-hand experience with formulating a worthwhile task, upper elementary students in South Korea "are provided with 'problem-posing'

---

[12] For example, as was shown in Chapter 3, Section 3.4, there is only one way to spend $15 to buy books priced $5 and $2, two ways to spend $25 to buy books of that price, and it is not possible to spend $15 (or $25) to buy books priced $4 and $2.

activities where they create new problems by changing conditions of the problem given" [Nam et al., 2013, p. 209]. At the same time, changing conditions of a problem is not always easy and the outcome must be skillfully examined from different directions. Indeed, one of the recommendations for teachers in Singapore to encourage metacognition (i.e., thinking about thinking) in the classroom points directly at posing non-routine mathematical problems – in order to build on students' knowledge, teachers need to know how to develop "more challenging tasks that stretch their thinking and deepen their understanding" [Ministry of Education Singapore, 2020, p. 18]. This know-how includes the appreciation of the notion of task's coherence.

6.2 Didactical coherence in problem posing

In order for technology to have a positive effect on problem posing and problem solving, one should not only know how to use it but, more importantly, how to interpret the results that a technology tool generates. In the context of problem posing, this interpretation requires the appreciation of what may be called the didactical coherence of a problem [Abramovich and Cho, 2008]. Didactical coherence consists of three independent components: numerical coherence, pedagogical coherence and contextual coherence. Briefly, the three coherences can be described as follows. Numerical coherence deals with a formal solvability of a problem. For example, whereas a problem of dividing twelve M&Ms among four children in a fair way is solvable (because 12 is divisible by 4), replacing 12 by 13 makes the problem unsolvable or, in other words, numerically incoherent (because 13 is not divisible by 4 and a single M&M cannot be divided into 4 equal pieces)[13]. Likewise, whereas it is possible to put 8 cookies on two plates of one type and three plates of another type so that the plates of each type have the same number of cookies (because $8 = 1 + 1 + 2 + 2 + 2$), it is not possible to similarly put

[13] Although the words division and divisibility sound similar, they have different mathematical meaning – the former is an operation, and the latter is a property. So, when checking whether one integer is divisible by another integer, we check the property that division does not yield a remainder. There is evidence that some elementary teacher candidates think of divisibility as an operation and when asked whether 13 is divisible by 4, might answer in affirmative using the equality $13 \div 4 = 3.25$ as a convincing justification.

nine cookies on three plates of one type and four plates of another type (see a formal explanation in Section 6.6 below). That is, increasing just by one the number of cookies and the number of plates of each type (towards developing a more "challenging" task) yields a numerically incoherent problem. One has to keep in mind that numeric data which makes word problems solvable may not be simply altered to yield a coherent problem-posing outcome. In other words, changing numeric conditions of a problem is not an easy task. Furthermore, as the above examples demonstrated, conceptual explanations of numerical coherence (incoherence) may have epistemically different makeup.

Pedagogical coherence of a problem deals with such issues as its grade-appropriateness, a method of solution expected, or the number of solutions the problem has. For example, whereas the problem of spending \$25 on buying books priced \$3 and \$4 has only two solutions (see the footnote to Section 6.1), the problem of spending \$20 on buying books priced \$5, \$2, and \$1 has 29 solutions and therefore, the problem seeking all the solutions in the absence of a (specifically designed) computational tool is not pedagogically coherent (see the spreadsheet of Fig. 6.1). Likewise, whereas the problem of putting 13 cookies on 5 plates of one type and 8 plates of another type has a single solution (with one cookie on each plate), there are nine ways to put 20 cookies on a single plate of one type and two plates of another type. Expecting (elementary) students without technology to find all twenty-nine or even nine correct answers does not represent a coherent problem-posing ramification.

Finally, contextual coherence (mentioned in passing at the end of Chapter 3) deals with context within which a word problem is formulated. For example, the contexts of buying (inexpensive) books or putting cookies on plates are friendly to young children. At the same time, the problems of a real estate auction or arrangement of leaves on a stem (leading to Fibonacci numbers) are uncommon for young children and, in other words, may not be considered contextually coherent for this population of students. As noted by mathematics educators in Canada, "when problem solving is ... relevant to students experiences and derived from their own problem posing, it furthers their understanding of mathematics and develops their math agency" [Ontario Ministry of Education, 2020, p. 75].

6.3 From systematic reasoning to a joint use of the modern-day tools

Solving problems through systematic reasoning is the next step on the journey from trial and error to logically rigorous solution. As noted by South African mathematics educators, "logic forms an integral part of mathematical thinking, however even adults sometimes find it difficult to reason in a formal logical way" [Department of Basic Education, 2018, p. 18]. The use of a spreadsheet can provide support in developing reasoning skills based on logic. As an illustration, consider the following problem.

It takes 51 cents in postage to mail a letter. A post office has stamps of denomination 5 cents, 7 cents, and 10 cents. How many sets of the stamps could Anna buy to send a birthday card to a friend?

The spreadsheet shown in Fig. 6.1 (its programming details are included in Chapter 13) displays the following information. In cells E1, F1, and G1 the denominations of available stamps are presented in the descending order. In cell A1 the required postage is displayed. The ranges [D3:J3] and [C4:C12] include possible quantities of the 10-cent and 7-cent stamps, respectively. In the (two-dimensional) range [D4:J12] the quantities of the corresponding 5-cent stamps (the smallest denomination) are generated by finding the value of c from the equation

$$10a + 7b + 5c = 51 \qquad (6.1)$$

where a, b and c ($0 \le a \le 5$, $0 \le b \le 7$, $0 \le c \le 10$) are possible quantities of the available stamps that make the required postage. One can see (range [D7:G7]) four values of c – the quantities of the smallest denomination stamp. For example, the triple (1, 3, 4) the elements of which are displayed in cells E3, C7, E7, respectively, indicates that with one 10-cent stamp, three 7-cent stamps, and four 5-cent stamps, the required postage, 51 cents, can be achieved. One can see that all four solutions include three 7-cent stamps. This observation points at some uniformity in the behavior of one of the variables that structure equation (6.1) and it can be used to explain how the problem may be solved by a student through systematic reasoning.

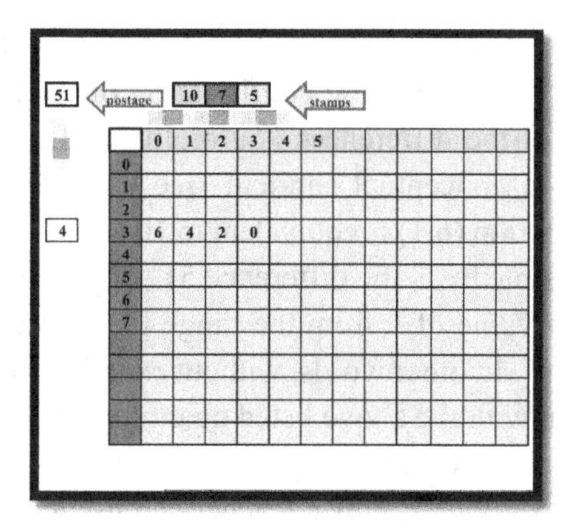

Fig. 6.1. Four ways to make 51-cent postage out of the three stamps.

Elementary students should have experience with systematic reasoning from early grades through decomposing a positive integer into a sum of two positive integers, a typical problem to appreciate the idea of a multiple representation [e.g., Becker and Selter, 1996; Van de Walle, 2001; Young-Loveridge, 2002; Serrazina and Rodrigues, 2015; Abramovich, 2021]. Therefore, the first question to be explored is whether, using only 10-cent and 5-cent stamps, the 51-cent postage can be made. Because 51 is not a multiple of 5 (alternatively, counting by fives would miss 51) and, therefore, it is not a multiple of ten, one cannot make this postage using 10-cent and 5-cent stamps only. This implies that one has to try using a 7-cent stamp also. Using a single 7-cent stamp requires one to add a 44-cent postage (51 – 7 = 44) using other two stamps, something that is not possible (because counting by fives misses 44 as well). Likewise, using two 7-cent stamps requires one to add a 37-cent postage (51 – 14 = 37) which is not possible using the other two stamps. Using three 7-cent stamps requires one to add a 30-cent postage (51 – 21 = 30) which, finally, can be made using 10-cent and 5-cent stamps only. Numerically, possible representations of 30 through 10 and 5 are:

$$30 = 10 + 10 + 10;\ 30 = 10 + 10 + 5 + 5;\ 30 = 10 + 5 + 5 + 5 + 5;$$
$$30 = 5 + 5 + 5 + 5 + 5 + 5.$$

These four representations, developed from the first one by trading a ten for two fives, correspond to the triples (3, 3, 0), (2, 3, 2), (1, 3, 4), and (0, 3, 6), in which the first, second and the third elements represent, respectively, the number of 10-cent, 7-cent and 5-cent stamps used to make the 51-cent postage Noting that $51 - 28 = 23$, $51 - 35 = 16$, $51 - 42 = 9$ and $51 - 49 = 2$, this numeric confirmation of the absence of other solutions (as none of the differences is a multiple of five) can be made more formal using a conceptual shortcut (see Chapter 3, Section 3.4). Indeed, because equation (6.1) is equivalent to $10a + 5c = 51 - 7b$ and both $10a$ and $5c$ are divisible by 5, the difference $51 - 7b$ has to be divisible by 5 as well. The only value of b from the range [0, 7] that makes $51 - 7b$ divisible by 5 is $b = 3$. In other words, without exactly three 7-cent stamps used in combination with the above listed quantities of 10-cent and 5-cent stamps, it is not possible to achieve the 51-cent postage.

This is an example of how a teacher can use technology not only to design a problem for their students but also to learn how to teach students to solve a problem through systematic reasoning (a method used to prove that all solutions have been found) in the absence of technology. Furthermore, technology can be used to "let students *actively contribute to the formulation* of the problem ... [keeping in mind that] the solution often needs less insight and originality than the formulation" [Pólya, 1981, p. 105, italics in the original][14]. Indeed, by changing the entries in the cells A1, E1, F1, and G1 and their contextual meaning, new problems can be formulated and easily solved within the spreadsheet. For example (see Fig. 12.2, Chapter 12, Section 12.2), using the spreadsheet of Fig. 6.1 one can find that there are 12 ways a quarter (cell A1: = 25) can be changed into dimes (cell E1: = 10), nickels (cell F1: = 5), and pennies (cell G1: = 1).

[14] A similar quote was presented at the end of Section 6.1.

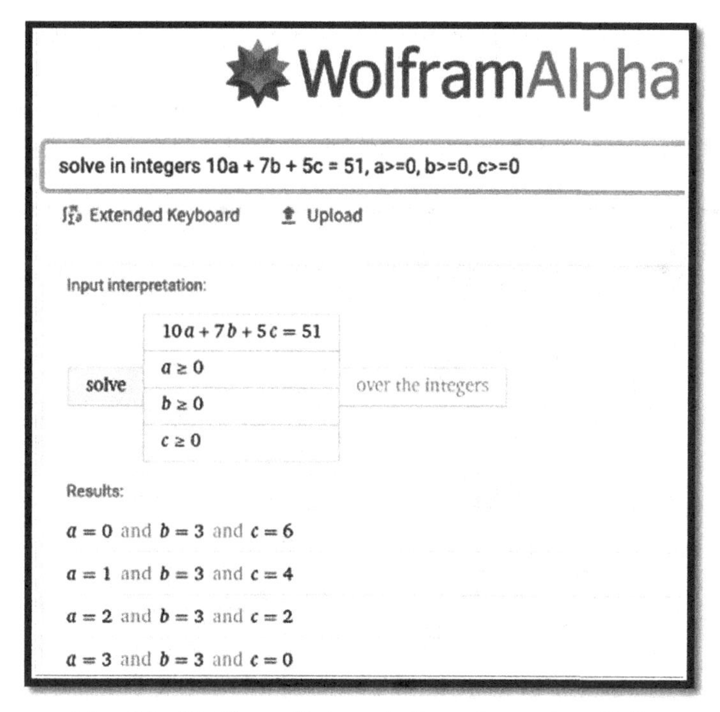

Fig. 6.2. *Wolfram Alpha* solution of equation (6.1).

Another digital tool capable of solving equation (6.1) with three unknowns is *Wolfram Alpha*. By entering into its input box the command "solve in integers 10a+7b+5c=51, a≥0,b≥0,c≥0" the solution to equation (6.1) follows (Fig. 6.2). One can see that *Wolfram Alpha* does not require any sophisticated programming except an accurate formulation of the task. Accurate formulation requires from a teacher to possess deep understanding of mathematics and its terminology. Deep understanding includes the ability to begin thinking about a problem in hand with some initial understanding of how its solution might look like [Arnheim, 1969]. At the same time, rather advanced knowledge of a spreadsheet makes it possible to extend explorations into equation (6.1) and not only find numerically the number of solutions of equation (6.1) for each value of its right-hand side but automatically keep records of such numbers in the form of a sequence generated in column S of the spreadsheet of Fig. 6.3. This sequence is interesting in itself, prompting connections to other tools of technology and making it possible to independently verify the accuracy of computations. Knowing the number of solutions to the postage-type tasks facilitates posing pedagogically coherent problems selecting not only the value of the right-hand side of equation (6.1) from column R of the spreadsheet of Fig. 6.3, but its coefficients as well.

	A	B	C	D	E	F	G	H	I	J	K	L	M	N	O	P	Q	R	S	T
1	20		postage		10	7	5	stamps												
2																		1	0	
3				0	1	2												2	0	
4			0	4	2	0												3	0	
5			1															4	0	
6			2															5	1	
7	3																	6	0	
8																		7	1	
9																		8	0	
10																		9	0	
11																		10	2	
12																		11	0	
13																		12	1	
14																		13	0	
15																		14	1	
16																		15	2	
17																		16	0	
18																		17	2	
19																		18	0	
20																		19	1	
21																		20	3	

Fig. 6.3. Recording the number of solutions of (6.1) as its right-hand side changes.

For example, the first value of the right-hand side responsible for three solutions (Fig. 6.3, cell S21) is the number 20 (Fig. 6.3, cell R21). As was shown in Fig. 6.2, the use of *Wolfram Alpha* provides an alternative verification that spreadsheet computations were correct. The next step is to use the Online Encyclopedia of Integer Sequences (OEIS®) to enter the first 20 terms, 0, 0, 0, 0, 1, 0, 1, 0, 0, 2, 0, 1, 0, 1, 2, 0, 2, 0, 1, 3, of the sequence generated by the spreadsheet in column S (Fig. 6.3). The OEIS® recognizes this sequence as the expansion of the fraction $1/[(1-x^{10})(1-x^{7})(1-x^{5})]$ which can be written in the form of the product of three sums of geometric series

$(1 + x^{10} + x^{20} + x^{30} + \ldots)\,(1 + x^7 + x^{14} + x^{21} + \ldots)\,(1 + x^5 + x^{10} + x^{15} + \ldots)$, so that, as discussed in [Abramovich and Brouwer, 2003; Abramovich, 2021], the coefficient in x^{51} in the expansion of this product is the number of solutions of equation (6.1). This coefficient can be provided by the OEIS®, see the 51st term of sequence A025884. The combination of three modern-day sources of information – an electronic spreadsheet, *Wolfram Alpha* and OEIS® – provides strong evidence of the accuracy of computations made possible by spreadsheet programming. In the words of Freudenthal [1978, p. 193], "it is independency of new experiments that enhances credibility".

6.4 From cookies on plates to spreadsheet modeling

Consider another contextual illustration of decomposing an integer into like addends: *How many ways can ten identical cookies be put on two plates with at least one cookie on a plate and by counting different orders of the plates?* Uncomplicated addition and its commutative property (to attend to the requirement of counting different orders of the plates) provide an answer to this question[15] through the following nine equalities (each describing one of the nine ways): $10 = 1 + 9 = 9 + 1 = 2 + 8 = 8 + 2 = 3 + 7 = 7 + 3 = 4 + 6 = 6 + 4 = 5 = 5$ (see Chapter 4, Section 4.1).

By extending this context, one can introduce the third plate filled with ten cookies and interpret the above nine decompositions of the number 10 as having three plates filled with cookies so that one of the plates has as many cookies as the other two plates combined. Adding a simplifying condition that plates are arranged in order (that is, ranked by ordinal numbers), one can still intricate the situation by extending it to four plates where the third plate has as many cookies as the first and the second

[15] The meaning of the assumption of cookies being identical is twofold. First, it shows how a simple question, in the absence of such an assumption, may require a mathematically complicated resolution, thus demonstrating an intricate relationship between context (from where real-life problem posing stems) and mathematics (which frames the ensuing problem solving). Second, in the case when such an assumption, while having been *tacitly* presumed, is questioned by a curious student, a teacher may simply say that the cookies are just cookies with their size, form, and type being immaterial (yet admitting that the student's query makes sense). Notwithstanding, in what follows, this assumption will be taken for granted.

plates combined, and the fourth plate has as many cookies as the second and the third plates combined. Note that adding the fourth plate makes the situation more complex in one sense and less complex in another sense for it leaves out the possibility of swapping the first two plates if the number of cookies on the fourth plate is specified. Now the following question can be posed:

If the fourth plate has ten cookies, how many cookies can one put on the first and the second plates, so that the third plate has as many cookies as the first two plates and the second and the third plates have ten cookies combined?

Fig. 6.4 shows a possible solution with two cookies on the first plate, four cookies on the second plate, six cookies on the third plate, and ten cookies on the fourth plate. In other words, the quadruple (2, 4, 6, 10) is such that $6 = 2 + 4$ and $10 = 4 + 6$. At the same time, swapping the first two numbers yields the quadruple (4, 2, 6, 8) which provides an erroneous response to the question posed and indicates the importance of distinguishing between the first two plates in terms of the number of cookies each plate has, given the number of cookies on the last plate. Below, we will refer to the process according to which cookies have been (and will be) put on plates as the ATLT (*add the last two*) rule. By reflecting on the solution shown in Fig. 6.4, one might wonder whether this solution is unique and if not, how all solutions to the problem with ten cookies on the fourth plate can be found. In a mean time, note that the quadruple (2, 4, 6, 10) can be extended to the quintuple (2, 2, 4, 6, 10), thereby, extending the problem to the fifth plate having ten cookies on it so that, according to the ATLT rule, $4 = 2 + 2$, $6 = 2 + 4$ and $10 = 4 + 6$.

By changing the number of cookies and the number of plates, new problems structured by the ATLT rule can be posed. Such changes should not be considered trivial – for example, one can check to see that whereas there is one way only to put seven cookies on the fifth plate following the ATLT rule, there is no solution for seven cookies and six plates. This is where technology may come into play under the umbrella of combined problem posing and solving, allowing for arithmetic, concrete materials and digital computation to meet.

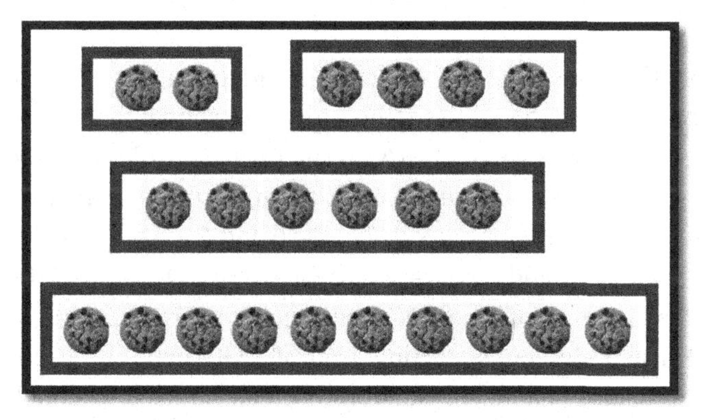

Fig. 6.4. One way to get 10 cookies on the fourth plate: (2, 4, 6, 10).

	A	B	C
1		10	4th plate
2	1st plate	2nd plate	
3	1		
4	2	4	
5	3		
6	4	3	
7	5		
8	6	2	
9	7		
9	8	1	
10			

Fig. 6.5. There are four ways to put 10 cookies on the fourth plate using the ATLT rule.

To begin, one can use a spreadsheet (Fig. 6.5) to find all possible arrangements of cookies on four plates satisfying the ATLT rule when the fourth plate has ten cookies (cell B1). To this end, in column A consecutive integers starting from one can be generated. Then, because, in general, the ATLT rule yields the sequence $(a, b, a + b, a + 2b)$, where the letters a and b stand, respectively, for the number of cookies on the first and the second plates, the equation

$$a + 2b = 10 \tag{6.2}$$

makes it possible to define the values of b satisfying this equation, given the value of a; that is, to express b through a. As shown in the spreadsheet of Fig. 6.5, the pairs (2, 4), (4, 3), (6, 2) and (8, 1) are solutions to equation (6.2) representing the quantities of cookies on the first two plates with ten cookies on the fourth plate. Indeed, $2 + 4 = 6$ and $6 + 4 = 10$; $4 + 3 = 7$ and $7 + 3 = 10$; $6 + 2 = 8$ and $8 + 2 = 10$; $8 + 1 = 9$ and $9 + 1 = 10$. In all four pairs, the first number of a pair is even ascending from 2 and the second number of a pair descends starting from 4. Alternatively, a conceptual shortcut approach (Chapter 3, Section 3.4) can be used to conclude that the value of a in equation (6.2) may only be an even number not greater than 8, as both 10 and $2b$ are even numbers. More specifically, there are four different solutions to the problem with ten cookies on the fourth plate developed through the ATLT rule: (2, 4, 6, 10), (4, 3, 7, 10), (6, 2, 8, 10), (8, 1, 9, 10). The computational solution can be verified by using cookies (or any concrete materials for that matter) through the creation of diagrams similar to Fig. 6.4, thereby doing a hands-on arithmetic.

Once the above four solutions to equation (6.2) have been found, another approach through the backward use of the ATLT rule can be discussed. Indeed, starting from 10, one can check to see if it can be preceded by 9. In that case, 9 is preceded by 1 and 1 is preceded by 8. Likewise, if 10 is preceded by 8, then 8 is preceded by 2 and 2 is preceded by 6. If 10 is preceded by 7, then 7 is preceded by 3 and 3 is preceded by 4. If 10 is preceded by 6 then 6 is preceded by 4 and 4 is preceded by 2. To demonstrate that all solutions to equation (6.2) have been found, one should note that if 10 is preceded by a number smaller than 6, then the next number is greater than 4 (=10 – 6) which makes it impossible using the ATLT rule, whatever the first non-zero number is. For example, if 10 is preceded by 5, then 5 has to be preceded by 5 (as $5 + 5 = 10$) and the only option for the first number is zero. Another way of reasoning is to decompose 10 into the sum of two (non-ordered) positive integers, $10 = 9 + 1 = 8 + 2 = 7 + 3 = 6 + 4 = 5 + 5$ to see that only the last decomposition contradicts the ATLT rule as it consists of two equal numbers implying that in order to reach 10 through this rule one has to start with zero (meaning no cookie on the first plate). At the same time, if the 5th plate has 10 cookies, then only the penultimate decomposition, $10 = 6 + 4$, does not contradict the ATLT rule as this is the only case when the 2nd plate has fewer cookies than the 3rd plate. This reasoning is a kind of a proof by the method of infinite descent or by the "well-ordering principle" [Apostol, 1967, p. 34] when, in the set of *positive* integers, the integer a precedes the integer b if and only if b is ether equal to a or is the sum of a and some positive integer. For example, the decomposition $10 = 7 + 3$ implies that $7 = 3 + 4$, which is impossible because if 4 precedes 3 then $3 = 4 + (-1)$ and the well-ordering principle is not satisfied.

This practice in formal reasoning is very helpful for it gives elementary teacher candidates much needed experience in "deep conceptual development in order to make sense of mathematics" [Department of Basic Education, 2018, p. 76]. Teacher candidates making sense of mathematics understand how to prepare students who are "mathematically proficient" [Common Core State Standards, 2010, p. 8], capable of "integrative thinking and problem solving" [Ontario Ministry of Education, 2020, p. 30], can "develop positive attitudes towards mathematics" [Ministry of Education Singapore, 2020, p. 22], and "solve problems by applying their mathematics to a variety of routine and non-routine problems with increasing sophistication" [Department for Education, 2013 updated 2021].

6.5 Spreadsheets, diagrams and number lines

It may be because of the self-contained nature of many uses of software tools in education, in general, and in mathematics education, in particular, negative affordances of technology have been often discussed. For example, some three decades ago, Blum and Niss [1991, p. 59] expressed a concern that the use of computers may reduce "intellectual efforts and activities of students... [to] mere button pressing". By the same token, Maddux and Johnson [2005, p. 2] claimed "the boring and mundane uses to which computers were often being applied has set the stage for a major backlash against bringing computers into schools". More recent research warned that instead of focusing on students' epistemic advancement through the use of digital tools, "teachers employ them to simply replace rather than transform existing approaches to instruction" [Bouygues, 2019, p. 8]. These, indeed, are legitimate concerns, and the key question is whether or not the benefits of the use of technology outweigh this risk. As in the above use of the spreadsheet of Fig. 6.5, it was used to perform quick, ATLT-based addition only, yet has offered little means of epistemic development. However, if one views technology as a tool for eliciting discovery rather than merely performing computation, it has to be viewed and used in a discovery fashion.

	A	B	C	D
1		7	4th plate	
2	1st plate	2nd plate		
3	1	3		
4	2			
5	3	2		
6	4			
7	5	1		
8	6			
9	7			

Fig. 6.6. Making the number of cookies on the 4th plate a variable.

For example, while the use of the spreadsheet of Fig. 6.5 is limited to four plates, it can be easily modified by making the right-hand side of equation (6.2) a variable to allow for the variation of the number of cookies on the fourth plate (Fig. 6.6). This modification would make it possible to see via computing that having both 8 and 7 cookies on the fourth plate yields three solutions, having both 10 and 9 cookies on the fourth plate yields four solutions, having both 12 and 11 cookies on the fourth plate yields five solutions, and so on. The following question can be posed (the correctness of an answer to which can be verified computationally using the spreadsheet of Fig. 6.6): *How many ways can cookies be put on the first two plates so that the use of the ATLT rule yields 35 cookies on the fourth plate*? This question can be generalized to n cookies. The answer is $\lceil n/2 \rceil - 1$ where $\lceil x \rceil$ is the smallest integer larger than or equal to x. Thus, $\lceil 35/2 \rceil - 1 = \lceil 17.5 \rceil - 1 = 18 - 1 = 17$. This answer can be verified computationally by using a spreadsheet which programming tools include a two-argument function $\lceil x \rceil = ceiling\,(x, 1)$ where the second argument means that x is rounded to the smallest integer greater than or equal to x.

In that way, one can generate a sequence of numbers by replacing the number 35 by any positive integer n. This sequence[16] looks like: 0, 0, 1, 1, 2, 2, 3, 3, 4, 4, 5, 5, ... For example, the rank of the second 4 in this sequence is 10 which means that there are four ways to put cookies on the first two plates so that the use of the ATLT rule yields 10 cookies on the fourth plate (as it was already shown in the spreadsheet of Fig. 6.5). Likewise, the rank of the first 3 in this sequence is 7 which means that there are three ways to put cookies on the first two plates so that the use of the ATLT rule yields 7 cookies on the fourth plate (as it was already shown in the spreadsheet of Fig. 6.6). Further, one can check to see that the term of this sequence with rank 35 (the number of cookies on the fourth plate) is equal to 17 (the number of ways cookies can be put on the first two plates). In the next section it will be demonstrated how a slight change in posing a problem, namely, by adding just an extra plate, significantly increases mathematical difficulty of the corresponding solution. In other words, a slight change in the condition of a problem can make a new problem pedagogically incoherent.

One can observe that an even number of cookies on the fourth plate requires an even number of cookies on the first plate and an odd number of cookies on the fourth plate requires an odd number of cookies on the first plate. How can this observation be explained? Having in mind that "observation and analogy may lead to discoveries" [Pólya, 1981, p. 158], note that due to the ATLT rule, one can provide an explanation using a very basic property of whole numbers: the sum of two numbers of the same parity (i.e., either both even or both odd) is an even number and the sum of two numbers of different parity is an odd number.

---

[16] The OEIS® includes this sequence under the number A004526 with many mathematical interpretations one of which is the number of partitions of a positive integer in two distinct (non-zero parts). Indeed, as was shown in Section 6.4, the plate with 10 cookies was created from the third and the second plates with, respectively, 9 and 1, 8 and 2, 7 and 3, 6 and 4 cookies. Thus, sequence A004526 has a situational referent in terms of cookies, plates, and the ATLT rule.

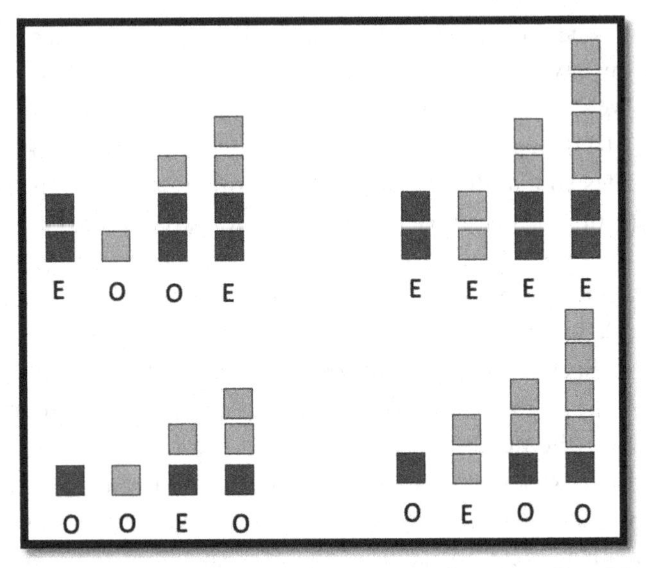

Fig. 6.7. The ATLT rule preserves the original parity of cookies for the fourth plate.

Fig. 6.7 provides a visual demonstration of the numerically observed phenomena: when starting with a pair (the smallest image of an even number), regardless, whether the pair is followed by a single or a pair, using the ATLT rule twice results in pairs; when starting with a single (the smallest image of an odd number), regardless, whether the single is followed by a single or a pair, using the ATLT rule twice does not result in pairs. Another observation is that the number of cookies on the fourth plate can be put on the first two plates as follows: once on the first plate and twice on the second plate (Fig. 6.8). This visual observation can be confirmed numerically:

$$10 = 2 + 4 + 4,\ 10 = 4 + 3 + 3,\ 10 = 6 + 2 + 2,\ 10 = 8 + 1 + 1.$$

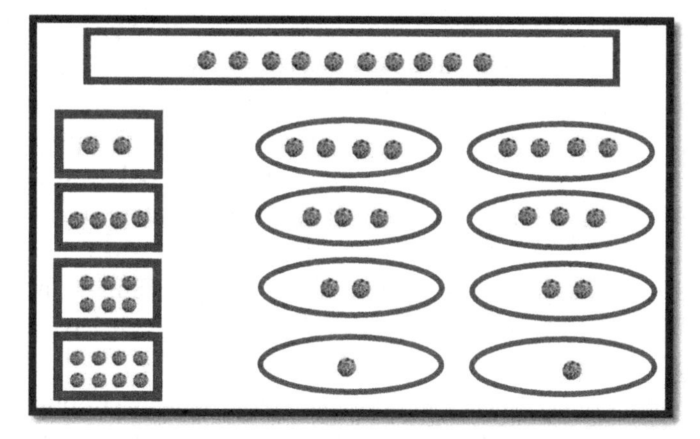

Fig. 6.8. Ten cookies on three plates of two types.

6.6 Posing new questions about cookies and plates

Extending the development of the diagrams of Fig. 6.7 through the ATLT rule results in the following four strings of letters E (even) and O (odd): EOOEOOEOOEOOE ..., EEEEEE ..., OOEOOEOOEOOEOOE ..., OEOOEOOEOOEOOE ... . That is, in a string of letters E and O, two consecutive E's are always separated by exactly two O's and the three-letter combination EOO in this string continues repeatedly (perhaps after a short transient period). This makes it possible to formulate the following questions about cookies and plates:

(i) If on the 10th plate there is an even number of cookies, is the number of cookies on the first plate even or odd?

(ii) If on the 100th plate there is an even number of cookies, is the number of cookies on the first plate even or odd?

(iii) If on the 10th plate there is an odd number of cookies, is the number of cookies on the first plate even or odd?

(iv) If on the 100th plate there is an odd number of cookies, is the number of cookies on the first plate even or odd?

The diagram of Fig. 6.9 constructed through the use of the above strings with two letters E and O provides answers to the above four questions. The number line at the top shows that plates of the ranks 1, 4, 7, ..., 100, ... , all have an even number of cookies (these rank numbers when divided by 3 yield the remainder 1). The number line in the middle shows that plates of the ranks 3, 6, 9, ..., 99, ... , all have even number of cookies (these rank numbers when divided by 3 yield the remainder 0) and therefore, the plates of the ranks 1, 10, and 100 have an odd number of cookies. Likewise, the number line at the bottom shows that plates of the ranks 2, 5, 8, ..., 101, ... , all have even number of cookies (these rank numbers when divided by 3 yield the remainder 2) and therefore, the plates of the ranks 1, 10, and 100 have an odd number of cookies. The case with an even number of cookies on the first two plates is not shown in Fig. 6.9 because, according to the ATLT rule, all the plates will have an even number of cookies.

Similar questions can be formulated for the plates of other ranks. To answer such questions computationally, a new spreadsheet environment will be created in the next section to accommodate the change in the number of plates involved. This would require a deeper inquiry into the ATLT rule.

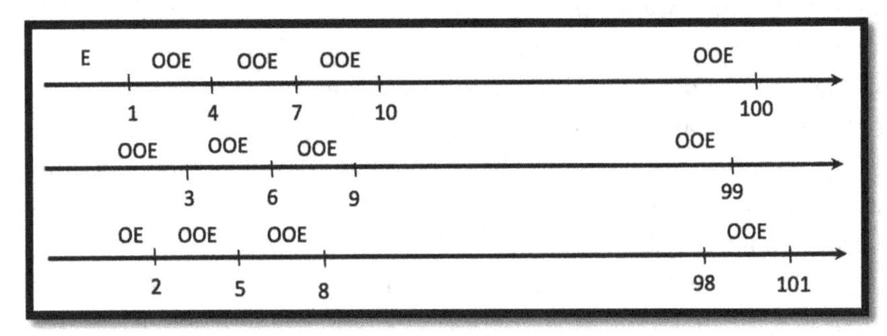

Fig. 6.9. A number line method of problem solving.

6.7 From cookies on plates to Fibonacci numbers

Fig. 6.8, partly informed by Fig. 6.5, shows four different ways to put ten cookies on three plates of two types (located horizontally), so that each triple of plates has either two or four cookies, four or three cookies, six or two cookies, and eight or one cookies. Using the W^{4}S (we write what we see) principle (Chapter 4, Section 4.1), one can describe numerically what is seen in Fig. 6.8 as follows: 10 = 2 + (4 + 4), 10 = 4 + (3 + 3), 10 = 6 + (2 + 2), 10 = 8 + (1 + 1). That is, all cookies on the fourth plate (i.e., ten cookies) can be put in three groups the cardinalities of which are either the number of cookies on the first plate or the number of cookies on the second plate and such arrangement of cookies can be done in four different ways One can see such three groups/plates in the four rows of Fig. 6.8 where each plate in the first row has either 2 or 4 cookies, each plate in the second row has either 4 or 3 cookies, each plate in the third row has either 6 or 2 cookies, and each plate in the fourth row have either 8 or 1 cookies.

Note that the computational environment of Fig. 6.6 can be modified to allow for a larger number of plates and cookies to be explored. Such a spreadsheet, the programming details of which are included in Chapter 13, is shown in Fig. 6.10. For example, the following problem can be posed:

How many ways can one put cookies on the first two plates so that, when using the ATLT rule, the fifth plate would have 25 cookies?

	A	B	C	D	E
1	cookies	25	plates	5	
2					
3	1st plate	2nd plate			
4	1				
5	2	7	9	16	25
6	3				
7	4				
8	5	5	10	15	25
9	6				
10	7				
11	8	3	11	14	25
12	9				
13	10				
14	11	1	12	13	25

Fig. 6.10. Four ways of ending up with 25 cookies on the 5th plate.

By entering (Fig. 6.10) in cell B1 the number of cookies at 25 and in cell D1 the number of plates at 5, the spreadsheet generates four quintuples (2, 7, 9, 16, 25), (5, 5, 10, 15, 25), (8, 3, 11, 14, 25), (11, 1, 12, 13, 25) as possible distribution of 25 cookies among 5 plates. One can check to see that using the first two elements of each of the four quintuples, the following representations of the number 25 result:

$25 = (11 + 11) + (1 + 1 + 1) = (8 + 8) + (3 + 3 + 3) = (5 + 5) + (5 + 5 + 5) = (2 + 2) + (7 + 7 + 7)$.

Such combination of concrete objects like cookies and digital tools like spreadsheets, being powerful in its simplicity, represents an engaging learning environment which motivates further explorations into the problems about cookies and plates. Together with pictorial representations of tactile-computational activities, this environment can "help to facilitate the study and communication of important mathematical results" [Ministry of Education Singapore, 2020, p. 8]. In particular, the spreadsheet of Fig. 6.10 can be used by a teacher to pose new problems for students to solve using concrete materials. For example, whereas a problem with seven cookies and four plates can be solved by trial and error (resulting in three solutions), the problem with 25 cookies and seven plates requires more than trial and error. This is where educational computing, can go beyond button pressing and, instead, support conceptual understanding and elicit epistemic advancement of the learners of mathematics.

Indeed, after several choices for the first two plates, one can start suspecting that the problem does not have a solution. A doubt calls for conceptualization which, in turn, requires generalization. Towards this end, the latter activity may begin, as before, with setting a and b to represent the quantity of cookies on the first and the second plates, respectively. Using the ATLT rule, one can develop the following sequence

$$a, b, a + b, a + 2b, 2a + 3b, 3a + 5b, 5a + 8b, \ldots \quad (6.3)$$

the displayed terms of which represent the number of cookies on the first seven plates, respectively. From here the equation

$$5a + 8b = 25 \quad (6.4)$$

follows describing 25 cookies on the seventh plate distributed in such a way that the first plate is repeated five times and the second plate is repeated eight times. One can attempt to solve equation (6.4) using a conceptual shortcut approach (Chapter 3, Section 3.4). Because both $5a$ and 25 are divisible by 5, the product $8b$ should be divisible by 5 also (otherwise, the equality between the left- and right-hand sides of equation (6.4) is not possible) implying that b is a multiple of 5. But already $b = 5$ – the smallest positive multiple of 5 – results in $8b = 40 > 25$. Therefore, indeed, the problem with 25 cookies and seven plates does not have a solution. This example shows the merit of generalization in understanding and resolving special cases. Without transition from arithmetic to algebra, that is, without replacing numbers by letters, equation (6.4) would never have come to light.

To explore the ATLT rule further, note that the coefficients in a and b in sequence (6.3) develop through the same rule. Indeed, in the sum $5a + 8b$ we have $5 = 3 + 2$ and $8 = 3 + 5$ where, likewise, $3 = 2 + 1$ and $2 = 1 + 1$. Mathematically, these coefficients can be defined as follows: its first two terms are equal to the number 1 and every term beginning from the third is the sum of the previous two terms. These numbers are called Fibonacci numbers[17]. Changing the first two Fibonacci numbers and keeping the rule according to which, all other terms develop, result in a sequence called Fibonacci-like sequence. A famous example of a Fibonacci-like sequence is represented by Lucas[18] numbers 2, 1, 3, 4, 7, 11, 18, 29, ... As shown in Chapter 13, Section 13.6, knowledge of the basics of Fibonacci numbers allows for the construction of the spreadsheet of Fig. 6.10.

Using the spreadsheet of Fig. 6.10, one can consider the case of five plates and then find the number of ways to put cookies on the first two plates so that the fifth plate has n cookies, $n \geq 5$. The first time this problem has a single solution is when $n = 5$. When $n = 6$ the problem does not have a solution. When $n = 7, 8, 9, 10$, the problem has a single solution defined, respectively, by the quintuples (2, 1, 3, 4, 7), (1, 2, 3, 5, 8), (3, 1, 4, 5, 9), (2, 2, 4, 6, 10). The problem with 11 cookies has two solutions defined by the quintuples (1, 3, 4, 7, 11), (4, 1, 5, 6,11). In that way, the sequence 1,

---

[17] The sequence of numbers 1, 1, 2, 3, 5, 8, 13, ..., nowadays commonly associated with the name Fibonacci, was also referred to as the series of Lamé, e.g., "la célèbre série de Lamé ou de *Fibonacci*" [Catalan, 1884, p. 8, italics in the original]. Gabriel Lamé (1795–1870) – a French mathematician and engineer. Eugène Charles Catalan (1814–1894) – a French/Belgian mathematician.

[18] Edouard Lucas (1842-1891) – a French mathematician, the author of the Tower of Hanoi puzzle [Hofstadter, 1985], who in 1876 gave the sequence 1, 1, 2, 3, 5, 8, 13, 21, ... its modern name [Koshy, 2001, p. 5].

0, 1, 1, 1, 1, 2, 1, 2, 2, 2, 2, 3, 2, 3, 3, 3, 3, ... (representing the number of ways to put cookies on the first two plates so that the fifth plate has n cookies, $n \geq 5$) can be developed[19]. One can see how adding extra plate significantly changed the complexity of the problem.

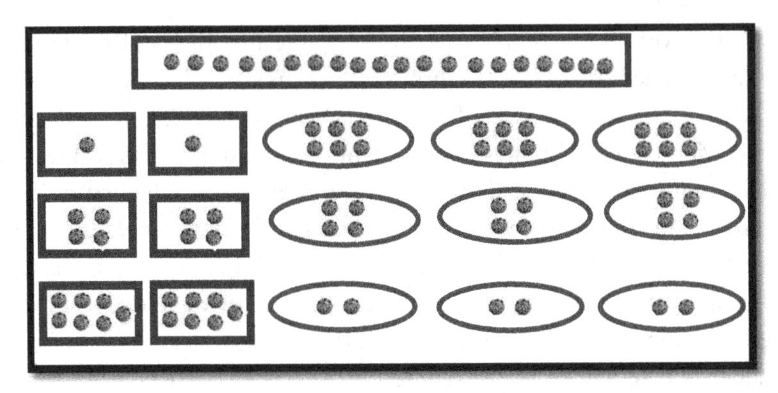

Fig. 6.11. Three ways to put twenty cookies on five plates of two types.

In order to put 20 cookies on the plates of two types (Fig. 6.11) – two plates of one type (rectangle) and three plates of another type (oval) – one can use the following systematic approach. If a cookie is put on each of the two plates of the first type, then the remaining 18 cookies can be put evenly on each of the three plates of another type (because 18 is divisible by 3). If two or three cookies are put on each of the two plates of the first type, then the remaining 16 or 14 cookies, respectively, cannot be put evenly on three plates of another type (because neither 16 nor 14 is divisible by 3). If four cookies are put on each of the two plates of the first type, then the remaining 12 cookies can be put evenly on three plates of another type (because 12 is divisible by 3). Continuing in the same vein, one can find that when seven cookies are put on each of the two plates of the first type, the remaining six cookies can be put evenly on the three

[19] The OEIS® includes this sequence under the number A103221 with many mathematical interpretations one of which is the number of partitions of a positive integer greater than 4 in parts 2 and 3 (each of which is a non-zero part). One can see that sequence A103221 has a situational referent in terms of cookies, plates, and the ATLT rule.

plates of another type (because 6 is divisible by 3). The arrangement of cookies is completed by noting that neither eight nor nine cookies put on the two plates of the first type allow for a solution in the context of the remaining three plates (because neither 4 nor 2 is divisible by 3).

6.8 From cookies to creatures to equations

Note that a problem with cookies can be reformulated using the context of pet store mathematics [Abramovich, 2005], keeping in mind that play-based activities serve as a motivating practice for young children allowing for "repetition and variation to achieve proficiency and flexibility" [Ministry of Education Singapore, 2020, p. 18]. One such problem can be formulated as follows:

A pet store sold two types of animals, drimps and grimps. It is known that among two drimps and three grimps there were twenty legs. How many legs does a drimp have and how many legs does a grimp have?

Here, the number of legs each creature has is the same as the number of cookies put on two plates of one type and three plates of another type. For example, Fig. 6.12, as a replica of Fig. 6.11, can be interpreted as drimps and grimps having, respectively, either 1 and 6 legs, or 7 and 2 legs, or both having 4 legs.

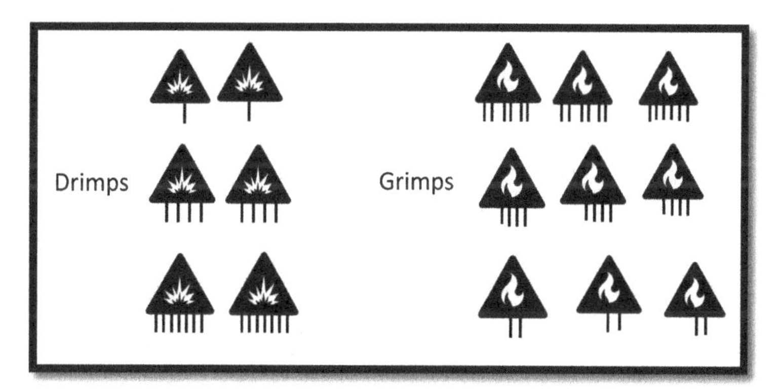

Fig. 6.12. Twenty legs among the creatures of two types (two drimps and three grimps).

This reformulation suggests a conceptual shortcut approach to solving the above problem. Note that the number 20, which is an even number, has to be represented as a sum of two numbers, one of which is even, and another is a multiple of three, which must also be even, otherwise the sum of two numbers would not be even. There are three even multiples of three smaller than twenty: 6, 12, and 18. Therefore, we have $20 = 14 + 6 = (7 + 7) + (2 + 2 + 2)$, $20 = 8 + 12 = (4 + 4) + (4 + 4 + 4)$ and $20 = 2 + 18 = (1 + 1) + (6 + 6 + 6)$. This approach can be formally described (through decontextualizing the diagrams of Fig. 6.12) as solving the equation $2D + 3G = 20$ by noting that because both $2D$ and 20 are divisible by 2, so is $3G$, something that is only possible when G is an even number. When $G = 2$ we have $D = 7$, when $G = 4$ we have $D = 4$ and when $G = 6$ we have $D = 1$. This completes the solution, as already when $G = 8$ we have $3G > 20$.

CHAPTER 7: FRACTIONS

7.1 From unit fractions to two contexts for fractions

The importance of unit fractions as building blocks of one's conceptualization of fractions and its arithmetic, is due to the appearance of the unit fractions long before the division, as a formal operation in the domain of whole numbers, is taught at the elementary level. (Not to mention here the focus of ancient Egyptians on the use of unit fractions discussed below in Section 7.8). Indeed, dividing something, a single thing, into equal parts is an early contextualized mathematical activity from which its decontextualization in the form of unit fractions stems. For example, in Singapore, primary school students are expected to "give examples of fractions in everyday situations, ... use concrete objects to represent and interpret fractions in terms of unit fractions" [Ministry of Education Singapore, 2012, p. 38]. In Japan "in grades 1 and 2, in order to understand fractions, fundamental mathematical activities such as folding a paper in half are carried out" [Takahashi et al.,2004, p. 13]. Furthermore, a (whole) pie can be divided in two, three, four, and so on *equal* (but not necessarily visually identical) parts and each part can then be given a mathematical name: one out of two equal parts (of the whole pie) can be called one-half (of the pie), one out of three equal parts (of the whole pie) can be called one-third (of the pie), one out of four equal parts (of the whole pie) can be called one-fourth (of the pie), and so on. These names can be assigned symbolic notations – 1/2, 1/3, 1/4, and so on, with the top number, 1, representing one whole (the pie) and the bottom number representing the number of equal pieces into which the whole is divided. Therefore, one can say that a whole is comprised of two halves, three thirds, four fourths, and so on. Just as symbolically, $\frac{1}{4}+\frac{1}{4}+\frac{1}{4}+\frac{1}{4}=1$, the sum $\frac{1}{4}+\frac{1}{4}+\frac{1}{4}$ can be described as $\frac{3}{4}$. This abstract symbol, a non-unit fraction, on a concrete level communicates two things: a certain whole (alternatively, "referent unit" [Conference Board of the Mathematical Sciences, 2012, p. 28]) has been divided in four equal parts (pieces) and three such pieces have been put together. In general, the fraction m/n, where m and n are positive whole numbers, can be seen as a numeric description of an outcome of putting together m unit fractions $1/n$. In that way, unit fractions provide a strong foundation for the study of non-unit fractions. However, arithmetical operations with fractions require the extension of integer arithmetic in the context of division to non-integer arithmetic. Similar to how arithmetical operations are motivated by the need to deal with large numbers (Chapter 2), the extension of integer arithmetic has been motivated by the need to deal with several division contexts leading to non-unit fractions. Note that the introduction of unit fractions and their use in constructing non-unit fractions occurs in the context that can be called part-whole – an easy-to-understand context which does not require any knowledge of arithmetical operations, division included. As the study of elementary mathematics progresses, the notion of the whole may also be extended to include more sophisticated context where a whole is comprised of more than one thing.

Remark 7.1. One can divide a rectangular pie (the whole) in two, three and four equal yet geometrically different parts (shapes) as shown in Fig. 7.1. Such division can first be demonstrated using a square grid so that both visually identical and distinct parts (fractions of the whole) can be compared by the number of unit squares they include (Fig. 7.1 (a, c, e)) followed by the cases when the areas occupied by one-thirds and one-fourths can be recognized as equal through drawing a bisecting diagonal (Fig. 7.1 (d, f)) or doing even more geometrically sophisticated partition in two equal parts like in the case of Fig. 7.1 (b), something that can be verified informally by using scissors. Note that in Fig. 7.1 each of the fractions 1/2, 1/3, and 1/4 has its own whole (that is, pies cut into two, three and four equal pieces differ by size). In upper elementary grades, the areas of the shapes shown in Fig 7.1 (b, d, f) can be found using Pick's formula (Chapter 10, Section 10.4), thereby connecting arithmetic, geometry, and discrete mathematics.

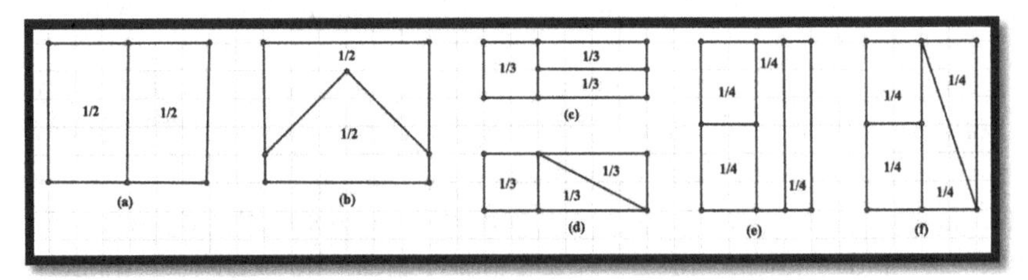

Fig. 7.1. Equal by size but geometrically distinct unit fractions.

There are two major real-life contexts leading to the concept of fraction as an extension of whole number arithmetic when the division of two whole numbers does not result in a whole number. One context, commonly referred to as part-whole, has its genesis in the early grades (as was discussed above); another is the dividend-divisor context, which by its name points at an upper elementary level. Recall that the operation of division was introduced through two modeling contexts: measurement context and partition context (Chapter 2, Section 2.4). One can say that the part-whole context for fractions extends the measurement context and the dividend-divisor context for fractions extends the partition context. To clarify the last statement, consider the relation which can be interpreted as measuring six donuts by two donuts as a way of creating three two-donut servings. Alternatively, this relation, that is, the division of six by two, can be interpreted as partitioning six donuts between two people evenly, allowing each person to receive three donuts. But what about dividing ten donuts among four people? While this is not impossible to do, fractions, typically, are introduced in the context of dividing a smaller number by a larger number. With this in mind, consider the case of dividing 3 by 4. As an extension of integer arithmetic, one can interpret this operation as measuring 3 donuts by 4 donuts resulting in a new type of number written as 3/4 which, in the form of *abstraction*, shows how many times the number 3 includes the number 4. Alternative interpretation of dividing 3 by 4 is partitioning 3 (identical) donuts among 4 people evenly which is more *concrete* than measuring 3 donuts by 4 donuts as the former activity would demonstrate physically what it means for each person to get 3/4 of a donut. Indeed, each of the three donuts can be divided in four parts and the resulting twelve pieces are then partitioned among four people so that each person gets three pieces of one-fourth of a donut. The former context of division (of 3 by 4) may be called the part-whole context where 3 is a part of 4 expressed through the number 3/4. The latter context of division (of 3 by 4) may be called the dividend-divisor context where 4 is a divisor of 3 resulting in the number 3/4.

Usually, teaching the part-whole context begins with dividing one thing into several equal parts. However, in integer arithmetic, teaching division as operation (Chapter 2, Section 2.4) does not begin with dividing the number one. Rather, in the context of measurement, we measure four by two, ten by five, twelve by three, and so on. And measuring three by four appears being beyond concreteness. In order to make this kind of measurement less abstract, the following clarification can be suggested. To begin, note that four donuts represent the unit of measurement using which three donuts have to be measured. In turn, this unit has to be represented through smaller units which are included both in four donuts and three donuts. One donut is such a unit, and it is one out of four smaller units of which the original unit – four donuts – is comprised. If one uses the notation 1/4 to describe one out of four donuts, then three donuts, when measured by that unit, can be described through the notation 3/4. In other words, whereas four donuts represent the whole, three donuts represent a part of the whole called, in a decontextualized form, three-fourths, or, symbolically, 3/4.

Often, the dividend-divisor context, although appearing as being less abstract, is not included into teaching fractions in the schools. A reason for such omission could be due to the fact that the part-whole context is already introduced through dividing an object (e.g., a rectangular-shaped pie) into several equal parts; for example, one pie divided in two (three, four) equal parts, each of which is called, respectively, one-half (one-third, one-fourth). These quantities are called unit fractions – one unit out of several into which the whole was divided. At the same time, it is important for elementary teachers to appreciate connection between fractions and division [Conference Board of Mathematical Sciences, 2012], something that "can be used to help bridge the gap between understanding fractions as parts of wholes and fractions as numbers" [Department of Basic Education, 2018, p. 37]. As far as unit fractions are concerned, it seems to be intuitively clear that unit fractions may be compared only when the whole that produces them is the same.

Sometimes, when talking about three (identical) pies being equally divided among four people (with three being the dividend and four being the divisor), instead of recognizing an opportunity for the dividend-divisor context to be introduced, a teacher might say in passing that each person would get 3/4 of a pie without paying attention to the significance of this context as perhaps the most natural way of extending integer arithmetic to that of fractions when dividend is smaller than divisor. Such cases explain the following assumption about mathematics teacher preparation: "Teaching mathematics effectively requires career-long learning ... [and] intentional efforts to seek additional knowledge ... [to be used] in supporting students' learning of mathematics" [Association of Mathematics Teacher Educators, 2017, p. 1].

The dividend-divisor context makes it possible to explain why a fraction has multiple representations through equivalent fractions and an integer has only one representation through itself. For example, $3/4 = 6/8 = 9/12 = 12/16 = \ldots$, and 3 has only one self-representation, $3 = 3$. But if one recognizes that in the fraction notation, a/b, the operation of division is hidden (that is, it is only identified, by presenting the dividend a and the divisor b, yet not completed), then at least two things become clear: there are many ways to represent three through an operation (e.g., $3 = 2 + 1 = 5 - 2 = 1 \times 3 = 12 \div 4$), and dividing three (identical) things in four equal parts gives the same result (i.e., the same quantity of material out of which the things are comprised) when six such things are divided into eight equal parts, nine – into twelve, and so on. Put another way, the dividend-divisor context makes it possible to introduce the notion of equivalence – a truly big idea of mathematics encountered by students at the elementary school level through the study of fractions. Likewise, dividing a pie (the whole) into four, eight, twelve, and sixteen equal pieces followed by taking three, six, nine, and twelve pieces, respectively, from the corresponding amounts bring about the notion of equivalence within the part-whole context as in each case the quantities of the pie are the same. One can see that from a pedagogical perspective, equivalence and invariance (Chapter 1, Section 1.1) – two big ideas of mathematics – require certain action on objects [Connell, 2001] to be involved.

Teacher candidates can be introduced to multiple representations of fractions, as a way of emphasizing the idea of equivalence, through the use of the so-called set model for fractions. The set model uses counters (or any identical objects) to represent fractions. The set model is considered more complicated than the other two models – area (using tape diagrams and paper folding) and measurement (using Cuisenaire[20] rods) models. The difficulty is in the use of multiple units in representing whole

[20] Georges Cuisenaire (1891-1975) – a Belgian primary school teacher.

numbers from which fractions develop. For example, if in the integer arithmetic, one counter represents the whole (a counting unit), in the fractional arithmetic, using a single counter as a whole does not allow for the development of a fraction. Let us represent a mixed fraction $1\frac{2}{3}$ through the set model. The first question to be answered is: what is the smallest number of counters that one can use to represent the whole? Because the fraction 2/3 has to be represented through counters as well, the whole should include three counters, two of which represent 2/3. The representation of $1\frac{2}{3}$ through the set model is shown in Fig. 7.2.

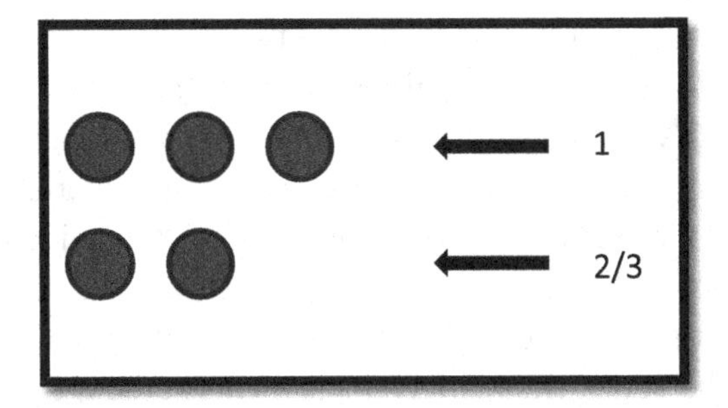

Fig. 7.2. The set model representation of $1\frac{2}{3}$.

But what if one has to subtract 1/6 from $1\frac{2}{3}$ using set model? In that case, Fig. 7.2 does not allow one to represent 1/6 and it has to be modified in order to demonstrate both $1\frac{2}{3}$ and 1/6. A modification requires the change in the number of counters representing the whole using the smallest number of counters. With this in mind, the whole has to be represented through six counters and 2/3 of this new whole would then be represented through four counters (Fig. 7.3). This allows one to carry out the subtraction as shown in Fig. 7.4. Put another way, six is the smallest number divisible by both three and six; in other words, six is the least common multiple of three and six.

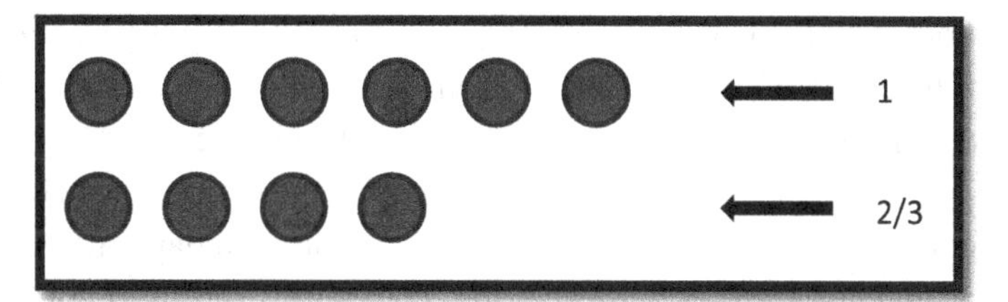

Fig. 7.3. A modified set model representation of $1\frac{2}{3}$.

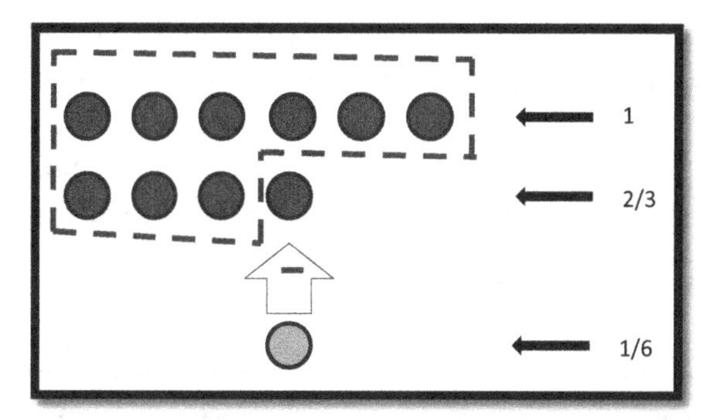

Fig. 7.4. The result of the subtraction is in the finger-pointer-like box.

There are multiple ways to represent numerically what we see in the finger-pointer-like box, provided that six counters represent the whole. One can see the nine counters as $1\frac{3}{6}$, or $1\frac{1}{2}$, or $\frac{9}{6}$, or $\frac{3}{2}$. That is, $1\frac{3}{6} = 1\frac{1}{2} = \frac{9}{6} = \frac{3}{2}$. This is an example of how conceptual understanding of equivalent fractions and, more generally, the idea of equivalence in mathematics can be developed and assimilated at the pre-operational level through the use of manipulatives.

Now, consider the fraction 3/5. One can introduce this fraction by extending partition model for division to non-integer arithmetic; for example, by dividing three identical objects such as pies among five people. As shown in Fig. 7.5, each of the three identical rectangles (representing pies) is divided into five equal parts. Through this process, fifteen equal size pieces result. Dividing 15 pieces among five people through the partition model for division yields three pieces of pie for each person, a piece being 1/5 of the whole pie. Thus, we have the fraction 3/5 representing the operation 3 ÷ 5, the outcome of which has only been identified through the notation 3/5, yet not completed. Connection between a fraction and division is mentioned by the Conference Board of the Mathematical Sciences [2012] as one of the most fundamental ideas in arithmetic that serves as a numeric characteristic of the ratio concept in middle school mathematics when the dividend-divisor comparison of the elements of two sets yields the same fraction. As was already mentioned, dividing three (identical) pies among five people results in the same quantity of a pie as dividing six such pies among ten people. That is, in both cases the ratio (see Chapter 9) of pies to people is the same, expressed through the fraction 3/5. At the same time, the fraction 3/5 can be understood within the context of dividing 15 pieces of pie among five people (through partition model for division) as repeated addition of the unit fraction 1/5 in the form $\frac{3}{5}=\frac{1}{5}+\frac{1}{5}+\frac{1}{5}$.

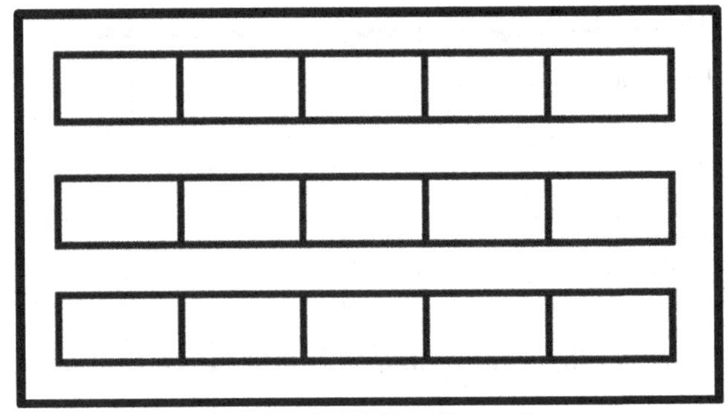

Fig. 7.5. Dividing 5 into 3.

The teaching of fractions conceptually can be enhanced by the use of the so-called tape diagrams [Common Core State Standards, 2010] aimed at explaining formal operations and their meaning. Just as the teaching of writing was recommended to "be arranged by shifting the child's activity from drawing things to drawing speech" [Vygotsky, 1978, p. 115], the teaching of arithmetic of fractions (including whole numbers) can be arranged as a transition from drawing a physical meaning of addition, subtraction, multiplication and division to describing the visual and the physical through culturally accepted mathematical notations. With this in mind, the W^4S principle (introduced earlier in Chapter 4, Section 4.1) can be used in the teaching of fractions. The diagram of Fig. 7.5 is a simple example of explaining the meaning of the fraction 3/5 by drawing an image of dividing three objects in five equal parts each. Note that the total number of pieces that this division provides is 15. It will be shown below (Section 7.9) how one can divide evenly three pies among five people having the number of pieces smaller than 15.

7.2 Adding and subtracting fractions

In order to develop skills in adding fractions, one can begin with reflecting on the addition of whole numbers. How do we add whole numbers within 10, i.e., before the base-ten system is introduced? For example, the meaning of the operation 3 + 2 stems from the context of adding three of something to two of something and the result is 5 (of something). But what if we add three pears and two bananas? The resulting number 5 represents neither pears nor bananas because the addends belong to different denominations. Is there a denomination to which both pears and bananas belong? The word fruit (or just thing) may be an answer to this question: adding three pears and two bananas yields five fruits (things). In the operation 3 + 2 both addends are abstractions, decontextualized from denominations they represent. But in this decontextualization, there is one concrete common characteristic – both 3 and 2 are comprised of the *same* unit of measurement. So, two units being added to three units yield five units. This unit is a fruit decontextualized from reality and abstracted to become a unit of measurement. Likewise, the equality 3 – 2 = 1 can be explained.

Similar situation is with fractions. Both 1/2 and 2/3, when subject to a common arithmetical operation, are *assumed* to be fractions (parts) of the same unit. However, the situation with fractions is more difficult than with whole numbers because the common denomination sought is both a part of the whole and a unit of measurement through which the fractions can be measured. So, talking about 1/2 and 2/3 of a pie (which is the unit), we have the two fractions of the pie to measure by another, fractional unit. In doing so, 1/2 and 2/3 of the pie are measured by 1/6 of the pie to have the total 7/6 of the pie. That is, decontextualization yields

$\frac{1}{2}=\frac{1}{6}+\frac{1}{6}+\frac{1}{6}=\frac{3}{6}$, $\frac{2}{3}=\frac{1}{3}+\frac{1}{3}=\frac{1}{6}+\frac{1}{6}+\frac{1}{6}+\frac{1}{6}=\frac{4}{6}$, and $\frac{1}{2}+\frac{2}{3}=\frac{7}{6}$. The same reasoning can be applied to the operation 2/3 – 1/2. Indeed, 2/3 – 1/2 = 4/6 – 3/6 = 1/6, where the notations 4/6 and 3/6 are understood as 1/6 repeated four and three times, respectively.

Four different types of fractions can be identified: a unit fraction (e.g., $\frac{1}{3}$ – a fraction with numerator one), a proper fraction (e.g., $\frac{2}{7}$ – a fraction with numerator smaller than denominator), an improper fraction (e.g., $\frac{7}{2}$ – a fraction with denominator smaller than numerator), and a mixed fraction (e.g., $5\frac{2}{3}$ but not $5\frac{2}{3}$). Indeed, $5\frac{2}{3}$ is a notation used to represent the sum of the integer part of a (rational) number and its fractional part, $5\frac{2}{3}=5+\frac{2}{3}$ (thus $5\frac{2}{3}$ is not a notation used to represent a mixed fraction as $\frac{3}{2}$ itself has a non-zero integer part). In order to turn a mixed fraction into an improper fraction, that is, to add an integer and a fraction, one has to find a common measure for the two numbers. To this end, one has to turn an integer into a fraction and add two fractions. In our case, $5\frac{2}{3}=5+\frac{2}{3}=\frac{5\cdot 3}{3}+\frac{2}{3}=\frac{5\cdot 3+2}{3}=\frac{17}{3}$. This explains conceptual meaning of a commonly known procedural rule of turning a mixed fraction into an improper fraction: $5\frac{2}{3}=\frac{5\cdot 3+2}{3}=\frac{17}{3}$.

The next two operations to be discussed in the context of fractions are multiplication and division. A common (proper or improper) fraction can be understood both from the perspectives of multiplication and division. Indeed, the fraction $\frac{a}{b}$ can be understood both as the unit fraction $\frac{1}{b}$ repeated a times and as the integer a divided by the integer b. Multiplication and division of fractions can be introduced by using area model for fractions, one of the three models for teaching and learning fractions already mentioned above: area, measurement, and set models. Conceptual meaning of the reduction of a fraction to the simplest form (something that was not observed in the context of whole number arithmetic) and the "invert and multiply" rule commonly utilized for the division of fractions will be discussed as well. A number of word problems that provide different real-life situations for modeling mathematics with fractions will be presented.

7.3 Multiplying two proper fractions

Where does the rule of multiplying two fractions come from? What is this rule? How, for example, can one justify, conceptually, that $\frac{2}{5}\cdot\frac{3}{7}=\frac{2\cdot 3}{5\cdot 7}=\frac{6}{35}$? More specifically, why does the operation of multiplication compel multiplying numerators and denominators of two fractional factors? The last question is consistent with expectations for students in the United States to "use the meaning of fractions, of multiplication and division ... to understand and explain why the procedures for ... fractions make sense" [Common Core State Standards, 2010, p. 33]. One can begin explanation from the multiplication of whole numbers where $2 \cdot 3$ is understood as repeating three objects twice, or, alternatively, taking two (identical) groups of any three objects. In the case of multiplying two fractions, repeating a fraction fractional number of times does not make sense unless this abstraction is put in context (contextualization) and explained through a two-dimensional diagram as shown in Fig. 7.6.

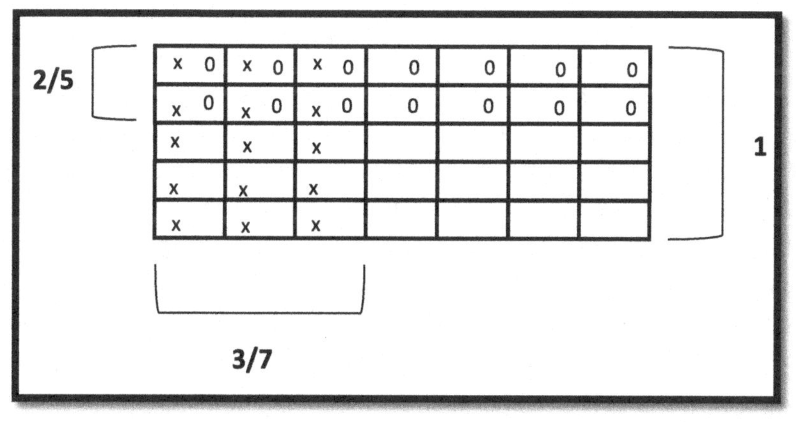

Fig. 7.6. Finding the product $\frac{2}{5}\cdot\frac{3}{7}$ within a grid.

Here, the large rectangular grid represents one whole, 3/7 of which are marked with x's and 2/5 of which are marked with 0's. In particular, by counting x's and 0's, the values of fractions 3/7 and 2/5 can be compared. Such counting is "helping students make sense of the operations and algorithms involving numbers" [Ministry of Education Singapore, 2020, p. 20].

Which part of the whole does the product represent? Could the product be represented by a region being larger than the whole? The answers to these questions are in the meaning of multiplication that does not change as the number system changes; that is, a physical meaning of multiplication is the same for objects described by whole numbers as for those described by fractions. When contextualizing the product $2 \cdot 3$, one points at the number of objects included in two groups *of* three objects. In other words, the multiplication sign is characterized contextually by the preposition 'of'; that is, multiplying two quantities means taking a certain quantity *of* another quantity. Likewise, $\frac{2}{5} \cdot \frac{3}{7}$ means taking $\frac{2}{5}$ of $\frac{3}{7}$; that is, taking the quantity 2/5 *of* the quantity 3/7. In order to make this operation more concrete, in other words, in order to demonstrate the skill of contextualization of the operation, one can first take $\frac{3}{7}$ of the whole (e.g., of a pie) and then take $\frac{2}{5}$ of the piece taken. This action results in a new fraction of the whole characterized by the product $\frac{2}{5} \cdot \frac{3}{7}$. This is a region where both marks, 0's and x's, overlap after 3/7 are marked with x's and 2/5 are marked with 0's. Now one can see the region described by the product. Yet, the question remains: how can one describe the region with both marks through a single fraction? The overlap of x's and 0's can be seen as 2 groups of 3 cells; that is, $2 \cdot 3 = 6$ cells belong to the overlap. The total number of cells in the whole can be seen as 5 groups of 7 cells; that is, $5 \cdot 7 = 35$ cells comprise the whole grid. Therefore, by moving from visual to symbolic, that is, through decontextualization, the overlap as a fraction of the grid (the whole) can be expressed numerically as $\frac{6}{35}$. In other words, $\frac{2}{5} \cdot \frac{3}{7} = \frac{2 \cdot 3}{5 \cdot 7} = \frac{6}{35}$. Alternatively,

$$\frac{2}{5}\cdot\frac{3}{7}=(\frac{1}{5}+\frac{1}{5})\cdot(\frac{1}{7}+\frac{1}{7}+\frac{1}{7})=\underbrace{\frac{1}{5}\cdot\frac{1}{7}+\ldots+\frac{1}{5}\cdot\frac{1}{7}}_{six\ times}=6\cdot(\frac{1}{5}\cdot\frac{1}{7}).$$

But contextually, the product $\frac{1}{5}\cdot\frac{1}{7}$ is understood as taking 1/5 of 1/7, something that, as shown in Fig. 7.7, is equal to 1/35. Thus,

$$6\cdot\left(\frac{1}{5}\cdot\frac{1}{7}\right)=6\frac{1}{35}=\underbrace{\frac{1}{35}+\frac{1}{35}+\ldots+\frac{1}{35}}_{six\ times}=\frac{6}{35}.$$

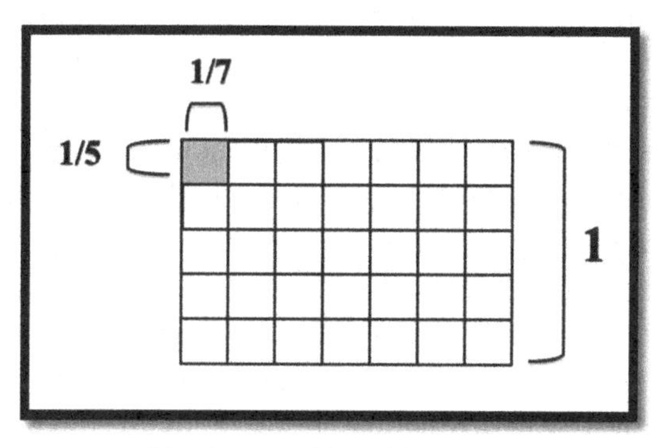

Fig. 7.7. Taking 1/5 of 1/7.

This approach to operations on fractions is based, in particular, on the expectation for students in England to "multiply proper fractions and mixed numbers, supported by materials and diagrams" [Department for Education, 2013, p. 34]. In general,

$$\frac{a}{b}\cdot\frac{c}{d}=\frac{a\cdot c}{b\cdot d}.$$

That is, the rule (algorithm) of multiplying fractions has been developed conceptually by assisting one in seeing "where a mathematical rule comes from" [Common Core State Standards, 2010, p. 4]. This position is consistent with a more recent statement of educators in the United States that "mathematics teaching ... requires not just general pedagogical skills but also content specific knowledge, skills, and dispositions" [Association of Mathematics Teacher Educators, 2017, p. 2]. These mathematics education positions are quite different from "early schooling practices ... [when the] "rule method" (memorize a rule, then practice using it) was the sole approach used in U.S. arithmetic textbooks from colonial times until the 1820s" [Conference Board of the Mathematical Sciences, 2012, p. 9].

7.4 Multiplying two improper fractions

In order to find out when the product of two fractions of a whole is represented by a region larger than the whole, consider the case of multiplying two improper fractions using a grid. How can one construct an image of the product $\frac{5}{2}\cdot\frac{7}{3}$ proceeding from an image representing one whole? As in the case of multiplying proper fractions, one can start with drawing a rectangle (the borders of which are solid lines) to represent the whole as shown in Fig. 7.8(a). The next step, shown in Fig. 7.8(b), is to divide (vertically) the whole into three equal parts, each of which is 1/3 of the whole and then extend it to the right by another four thirds to get 7/3. Then, as shown in Fig. 7.8(c) the whole is divided (horizontally) in two equal parts, each of which is 1/2 of the whole, and then extended down by another three halves to get 5/2 of the whole. As a result, 15 cells have been marked with x's. Now, the diagram of Fig. 7.8(b), representing the fraction 7/3, is divided in two equal parts resulting in 14 cells marked with 0's as shown in Fig. 7.8(d). Finally, taking 5/2 of 7/3 by overlaying the (c) and (d) parts of Fig. 7.8, yields the product $\frac{5}{2}\cdot\frac{7}{3}$ shown in Fig. 7.8(e).

Furthermore, it follows from Fig. 7.8(e) that the cells marked with x's alone represent 3/2 of the whole, the cells marked with 0's alone represent 4/3 of the whole, the cells with both marks represent the whole, and the cells with no marks represent 3/2 of 4/3. Put another way, the diagram of Fig. 7.8(e) reflects the distributive property of multiplication over addition:

$$\frac{5}{2}\cdot\frac{7}{3}=\left(1+\frac{3}{2}\right)\left(1+\frac{4}{3}\right)=1+\frac{3}{2}+\frac{4}{3}+\frac{3}{2}\cdot\frac{4}{3}.$$

Note that unlike the case of the product of two proper fractions, when their overlap is their product located within the whole, the overlap of the product of two improper fractions, as shown in Fig. 7.8(e), is the whole itself. Furthermore, the equality $\frac{5}{2}\cdot\frac{7}{3}=\frac{35}{6}$ means that the product consists of 35 cells each of which has a name of the unit fraction 1/6 given by the whole 6/6, is comprised of six such cells.

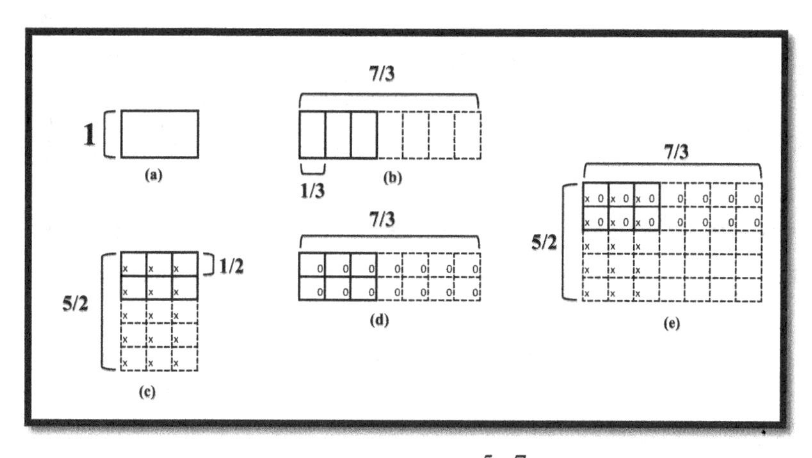

Fig. 7.8. From the whole to the product $\frac{5}{2}\cdot\frac{7}{3}$ as the sum of four non-overlapping regions.

Likewise, the product $\frac{5}{2}\cdot\frac{3}{7}$ in which only the first factor is an improper fraction is represented by a region that is larger than one whole. Of course, not every product of a proper and an improper fraction is greater than one. For example, the product $\frac{5}{2}\cdot\frac{2}{7}$ is smaller than one. Whereas the product of two proper or two improper fractions is, respectively, smaller or greater than the whole, the product of two fractions of different kind may be greater or smaller than the whole. One can see (Fig. 7.9) that the product $\frac{5}{2}\cdot\frac{3}{7}=\frac{3}{7}\cdot\frac{5}{2}$ (it is easier to see how one can take 3/7 of 5/2) is the sum of two non-overlapping regions: the regions with both x's and 0's and the region immediately below it with 0's alone.

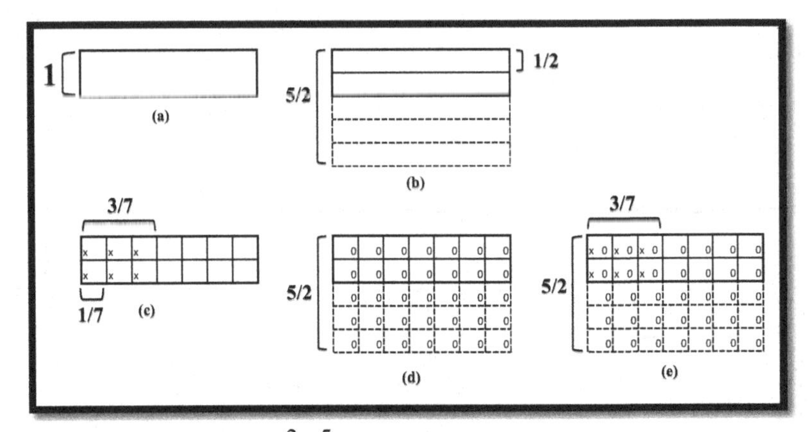

Fig. 7.9. The product $\frac{3}{7}\cdot\frac{5}{2}$ as the sum of the two non-overlapping regions: $\frac{6}{14}$ and $\frac{9}{14}$.

7.5 Dividing fractions

Multiplication and division (just as addition and subtraction) are operations in the relation of reciprocity. As was already observed with whole numbers, division is a more complicated operation than multiplication. The major difference between the two operations is that whereas the product of two whole numbers is a whole number, their quotient may bring with it a non-zero remainder. However, the set of rational numbers (that includes negative numbers not discussed here) is closed for all four arithmetical operations, except for the case of a zero divisor. So, both the product and the quotient of two fractions are fractions (not including division by zero). The goal of this section is to demonstrate how division of fractions can be carried out conceptually using (two-dimensional) area model for fractions just as it was carried out when multiplying fractions.

7.5.1 Dividing whole numbers

To begin, consider the case of dividing whole numbers by using area model. What does it mean to divide 5 into 4 (alternatively, dividing 4 by 5)? Because $5 > 4$, the smaller number cannot be measured by the larger number as one of the interpretations of division suggests (unless we develop an extended interpretation of the division of whole numbers). That is, in terms of the measurement model for division, the inclusion of 5 into 4 is an abstraction, as something larger may not be physically included in its part. Nonetheless, this abstraction can be represented in the form of a (positive) fraction smaller than one. This fraction also stems from the partition model for division in the context of dividing 4 by 5, something that, through an appropriate contextualization seems to be more concrete. The partition model for division in this context means dividing four things among five people in a fair way. Obviously, one should be able to cut things into equal pieces; that is, the things to be cut should be conducive to be partitioned in smaller (equal) parts. For example, it is not possible to divide four marbles in five equal parts. Yet, it is possible to make five servings from four (identical) pies. To do that, one has to divide each pie into five equal parts, so that a serving is 4/5 of a pie. Put another way, in order to find how many times 5 is included into 4 one has to divide 4 by 5. (In particular, this shows connection between fractions and division). How can one carry out this division physically (e.g., using a picture)?

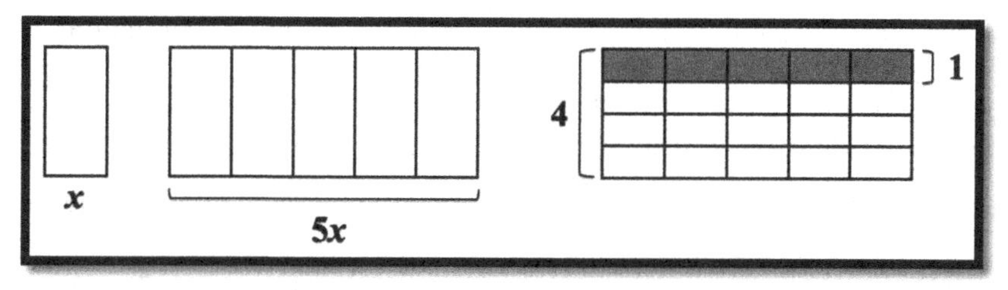

Fig. 7.10. Using two-dimensional model when dividing 5 into 4.

Let $4 \div 5 = x$. Then x is a missing factor in the equation $5x = 4$. The far-left part of Fig. 7.10 shows x. The middle part of Fig. 7.10 shows the left-hand side of the last equation, that is, $5x$. At the same time, the far-right part of Fig. 7.10 represents the right-hand side of the equation, that is, four horizontally arranged units. Consequently, the (referent) unit is shaded dark and it is one-fourth of the middle part of Fig. 7.10 (i.e., 1/4 of $5x$). The two-dimensional model shows that because the unit consists of five cells, each of which represents the fraction 1/5, previously unknown x becomes known as it consists of four such cells (vertically arranged and shown in the far-right part of Fig. 7.10). That is, $x = \frac{4}{5}$ or $4 \div 5 = \frac{4}{5}$. In other words, 4/5 is a number the repetition of which 5 times yields the number 4 (see Chapter 2, Section 2.4, the first paragraph).

Remark 7.2. One can describe the process of dividing 5 into 4 in terms of the change of unit. Indeed, selecting x as the original unit leads to the creation of $5x$ as shown in Fig. 7.10. Equating $5x$ to 4 suggests designating 1/4 of 4 as a new unit which, as shown in the far-right part of Fig. 7.10, consists of five cells. These five cells (as a new unit) can be expressed through x as $5x/4$. That is, in the process of dividing 5 into 4, the original unit is x and the new unit is $5x/4$. One can see that the latter turns into the number 1 when $x = 4/5$ – the result of dividing 5 into 4.

7.5.2 Dividing proper fractions

Now, using the same method, let us divide two proper fractions. For example, let us find the value of $\frac{4}{5} \div \frac{3}{4}$. The result of this division is a number x, three-fourth of which is equal to four-fifth; that is, x is the missing factor[21] in the equation $\frac{3}{4}x = \frac{4}{5}$. Fig. 7.11 includes four diagrams, the far left one, (a), representing x. The next diagram, (b), shows that x consists of four equal sections (towers) with an unknown numerical value, three of which, representing $\frac{3}{4}x$, are shaded. In order to make x known, one has to define the unit. To this end, because $\frac{3}{4}x$ is equal to $\frac{4}{5}$ (of the unit), the shaded part (along with the entire x) is divided horizontally into four equal parts so that one such part represents 1/5 of the unit. Therefore, the shaded area has to be extended down by 1/5 in order to show the unit, 5/5. This extension is shown in Fig. 7.11(c) using dotted lines. Through this process, the unit turned out being divided into $3 \cdot 5 = 15$ equal sections each of which is 1/15 of the unit. At the same time, we see that x comprises $4 \cdot 4 = 16$ such sections. That is, $x = \frac{16}{15}$. Put another way, as shown in

---

[21] In abstract terms, this missing factor shows which number has to be repeated 3/4 times to get 4/5 (compare the case of equation $3x = 12$ (see Chapter 2, Section 2.4) when the missing factor x stands for the number to be repeated 3 times to get 12).

Fig. 7.11(d) where dotted lines are the borders of the unit, in order to divide $\frac{4}{5}$ by $\frac{3}{4}$ one has to multiply $\frac{4}{5}$ by the reciprocal of $\frac{3}{4}$; that is, to take 4/5 of 4/3 (see Fig. 7.11). By using x's and 0's to mark cells that belong, respectively, to 4/3 and 4/5, one has the overlap of both marks coinciding with the unknown x. Numerically, we have $\frac{4}{5} \div \frac{3}{4} = \frac{4}{5} \cdot \frac{4}{3} = \frac{16}{15}$. Below, the meaning of this rule, commonly referred to as *Invert and Multiply*, will be explained in a different way.

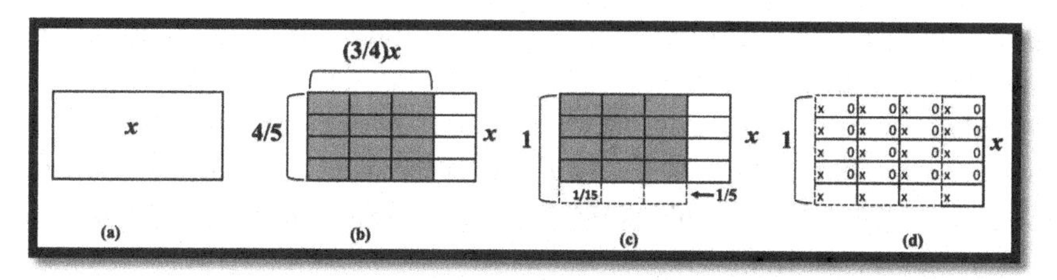

Fig. 7.11. Using two-dimensional method when dividing 4/5 by 3/4.

7.6 Conceptual meaning of the *Invert and Multiply* rule

Using the *Invert and Multiply* rule allows one instead of dividing two fractions to multiply the dividend and the reciprocal of the divisor. This rule makes it possible to divide the product of two whole numbers by another such product. For example, in the case $\frac{4}{7} \div \frac{3}{5}$ it might be tempting, by analogy with multiplying two fractions when their numerators and denominators are multiplied, to proceed dividing numerators and denominators to have $\frac{4 \div 3}{7 \div 5}$. This, however, brings us back to the division of the original two fractions by using the dividend-divisor context for division in the case of fractions. The *Invert and Multiply* rule allows one not to deal with the fraction $\frac{4 \div 3}{7 \div 5}$, but rather to compute the fraction $\frac{4 \cdot 5}{7 \cdot 3}$ in which (integer) numerators are multiplied by (integer) denominators.

7.6.1 The case of dividing whole numbers

In order to make sense of the conceptual meaning of the *Invert and Multiply* rule, one has to see how it works in the context of the division of two whole numbers. For example, the division fact $8 \div 4 = 2$ can be interpreted as finding how many 4-tile towers can be built out of 8 tiles (Fig. 7.12). Here, the numbers 4 and 8 point at a tile as the unit, and, therefore, measuring 8 tiles by 4 tiles in the process of building towers yields 2 towers. If division is replaced by multiplication and 4 is replaced by 1/4 (the reciprocal, or inversion, of 4), we have the relation $8 \cdot \frac{1}{4} = 2$. This time, we have eight one-fourths; that is, tile is not the unit anymore, but it is 1/4 of a tower, which became a new unit through using the *Invert and Multiply* rule. Repeating one-fourth of a tower eight times yields two towers. That is, the *Invert and Multiply* rule makes sense in the case of division of two whole numbers when the measurement model is used.

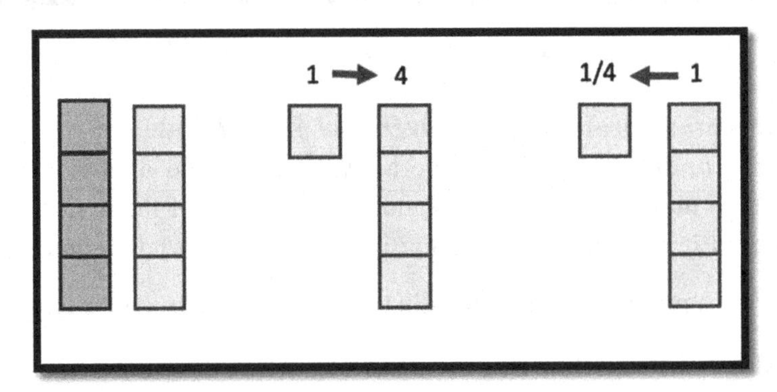

Fig. 7.12. Building four-tile towers out of eight tiles: $8 \div 4 = 8 \cdot \frac{1}{4}$.

However, if one has to build four identical towers out of eight tiles, we also have a process described as $8 \div 4 = 2$ but, this time, we use partition model for division. Replacing $8 \div 4$ by $8 \cdot \frac{1}{4}$ does not result in the change of unit as 8 and 4 are comprised of different units – tiles and towers, respectively. Due to the commutative property of multiplication, $8 \cdot \frac{1}{4} = \frac{1}{4} \cdot 8$ and, therefore, $8 \div 4$ is replaced by taking one-fourth of eight tiles. This yields two tiles for a tower. So, in the case of partition model for division the change of unit that happens through the use of *Invert and Multiply* rule is hidden and can be explained in terms of reducing the division equation $8 \div 4 = x$ to that of $4x = 8$, the solution of which is shown in Fig. 7.13. That is, in much the same way as described above in Remark 7.2, selecting x as the original unit allows one to create $4x$ and equate it to 8. This suggests that the new unit is 1/8 of the 8 and, it consists of four cells (the far-right part of Fig. 7.13). Those four cells can be expressed through the original unit x as $x/2$. One can see that the latter turns into the number 1 when $x = 2$ – the result of dividing 8 by 4.

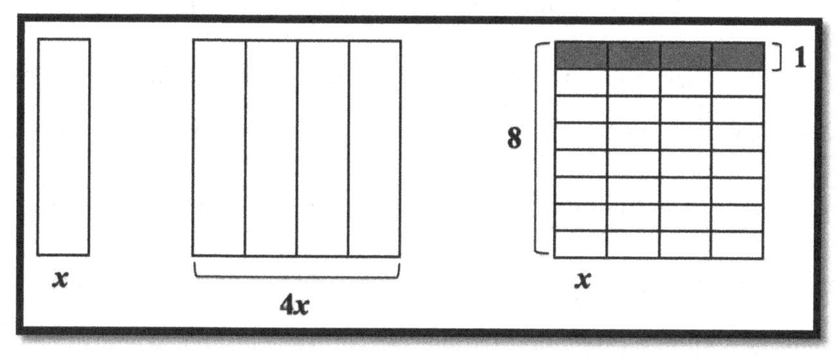

Fig. 7.13. Dividing 8 by 4 through the *Invert and Multiply* rule as a change of unit.

7.6.2 The case of dividing fractions

To begin, consider the case with the divisor is a unit fraction. Where can one recognize in the *Invert and Multiply* rule the change of unit when making a transition from the division $\frac{3}{4} \div \frac{1}{3}$ to the multiplication $\frac{3}{4} \cdot 3$ according to this rule? Which one is the original unit (in the case of division) and which one is the new unit (in the case of multiplication)? To answer these questions, let $\frac{3}{4} \div \frac{1}{3} = x$. Then $\frac{1}{3} \cdot x = \frac{3}{4}$. The first fraction in the last relation is 1/3. One may assume that x is the original unit in the problem with a missing factor. This unit, x, and its one-third are shown in Fig. 7.14, parts (a) and (b), respectively. Therefore, the new unit is the region (tower) for which $\frac{1}{3}x$ is equal to $\frac{3}{4}$. This new unit is shown in

Fig. 7.14 (c), the non-shaded part of which is 1/4 and the shaded part is 3/4. Then, the found value of $x = 3 \cdot \frac{3}{4} = \frac{9}{4}$ is shown in Fig. 7.14 (d).

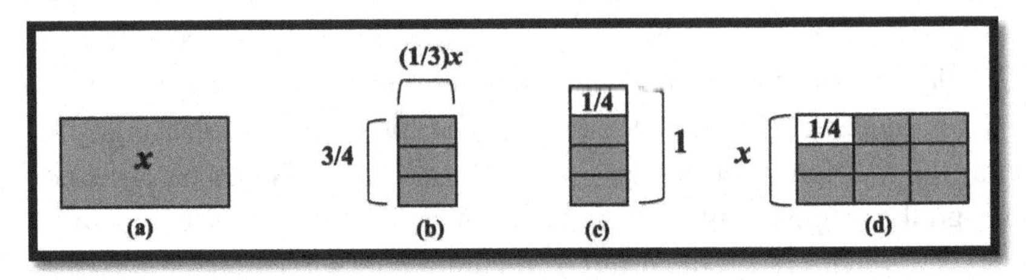

Fig. 7.14. From x being a unit to $4x/9$ being a unit.

In order to explain the *Invert and Multiply* rule in the context of mixed fractions, consider the division $3\frac{1}{3} \div \frac{5}{6}$. The operation can be contextualized and interpreted as measuring the distance of $3\frac{1}{3}$ miles by the segment of length $\frac{5}{6}$ miles when a truck covers any segment of this length in a minute (assuming the uniform movement; that is, movement with a constant speed throughout the entire distance; see Chapter 9, Section 9.5.2). This measuring would result in the number of minutes the truck needs to cover this distance. Here, one mile is the unit, and the truck makes 5/6 of the unit per minute. Replacing division by multiplication and the fraction 5/6 by its reciprocal, 6/5, yield $3\frac{1}{3} \cdot \frac{6}{5} = \frac{10 \cdot 6}{3 \cdot 5} = 4$. This time, the unit is the segment ran by the truck in a minute; this segment is smaller than 1 mile. Through this process, 1 mile becomes 6/5 of the segment and the later becomes a new unit (Fig. 7.15). If mile is a unit, then the segment, run in a minute, is 5/6 (miles). If a segment is a unit, then one mile is 6/5 (of the segment). Having $3\frac{1}{3}$ miles means that 6/5 of this segment (that is, 1 mile) when repeated $3\frac{1}{3}\left(=\frac{10}{3}\right)$ times yields 4 minutes. Note that in the described case, the measurement model for division was used.

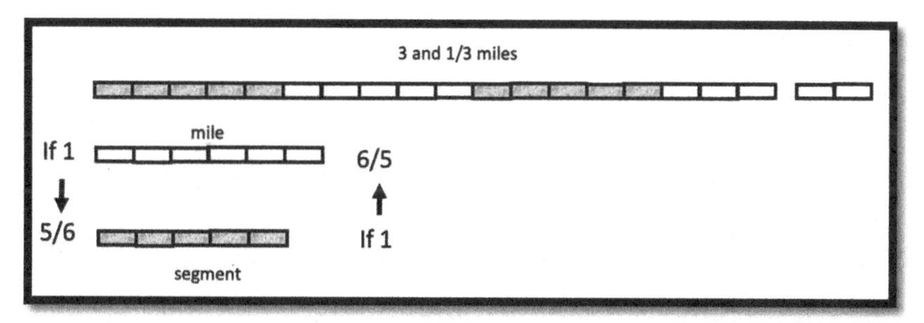

Fig. 7.15. From mile as a unit to its part as a unit.

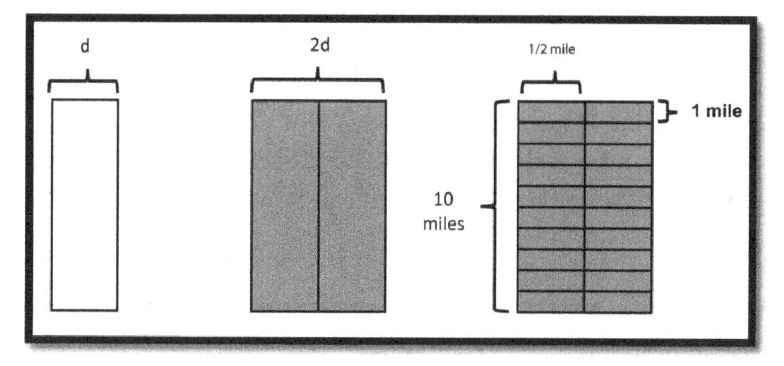

Fig. 7.16. From *d* as a unit to mile as a unit.

The situation is more difficult in the case of division used as an operation needed to find a missing factor when another factor and the product are given. For example, when twice the distance (d) from A to B is 10 miles, we have the equation $2d = 10$ whence $d = 10 \div 2 = 5$. This time, just as in the case of building four-tile towers out of eight tiles, the meaning of the division operation used to find d is distributing 10 miles within 2 distances (this distribution is the partition model for division). The result is 5 miles for the distance which is the unit. The process of finding the missing factor d from the equation $2d = 10$ can be shown on a picture. The far-left and the middle parts of Fig. 7.16 represent, respectively, the distance d and its double, $2d$. Distributing 10 miles among two distances through partition model for division is shown in the far-right part of Fig. 7.16. One can see that in the operation $2d$ we have distance d as the original unit. The operation $2d$, replacing division by multiplication and inverting the divisor 2, points at the fraction 1/2 as one-half of a mile. That is, through the *Invert and Multiply* rule the distance d as the original unit was replaced by a mile as a new unit, so that $d = 5$ miles.

Consider now the situation when 5/9 of unknown distance d is equal to 20/3 miles. Finding d from the equation $\frac{5}{9}d = \frac{20}{3}$ leads to the division $d = \frac{20}{3} \div \frac{5}{9}$. This time, the process of distributing 20/3 miles within 5/9 distances (d) is difficult to imagine[22] for it is an abstraction; yet the distance d may be the unit when the operation $\frac{20}{3} \div \frac{5}{9}$ is considered. Replacing $\frac{20}{3} \div \frac{5}{9}$ by $\frac{20}{3} \cdot \frac{9}{5}$ and the latter by $\frac{9}{5} \div \frac{20}{3}$ means the change of unit; that is, the distance d becomes 9/5 of 20/3 miles, which is 12 miles, where one mile is the new unit. The process of finding the missing factor d from the equation $(5/9)d = 20/3$ can be visualized using a picture (Fig. 7.17).

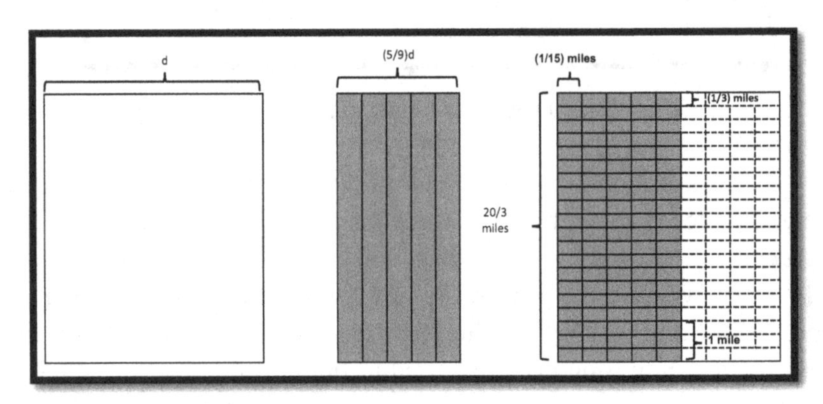

Fig. 7.17. From d as a unit to mile as a unit.

[22] One can replace 20/3 miles by 60/9 miles; then, distributing 60/9 miles among 5/9 distances may be seen as distributing (fairly) 60 miles among 5 distances which results in 12 miles for a distance. This suggests that instead of using the *Invert and Multiply* rule, one can proceed as follows: (20/3) ÷ (5/9) = (60/9) ÷ (5/9) = (60 ÷ 5)/(9 ÷ 9)= 12/1 = 12. One can see such an alternative to the *Invert and Multiply* rule as a conceptual shortcut (Chapter 3).

The far left and the middle parts of Fig. 7.17 represent, respectively, the distance d and $(5/9)d$. The far-right part of Fig. 7.17 shows how 20/3 miles can be distributed among $(5/9)d$, each row being 1/3 of a mile so that three such rows being one mile. One can see that there are 15 small boxes in one mile, each box representing 1/15 of a mile. Finally, because there are $9 \cdot 20 = 180$ such boxes in the large box representing d, we have $180(1/15) = 12$ miles. That is, through the *Invert and Multiply* rule the distance d as a unit was replaced by a mile as a new unit. The meaning of operations in the case of fractions with a missing factor can only be explained by referring to a similar case that involves whole numbers (see Fig. 7.13 and the case discussed in Section 7.5.1); otherwise, the partition model for division, unlike the measurement model for division, is an abstraction.

Remark 7.3. One may notice that procedurally, adding or subtracting proper (or improper) fractions is more difficult than multiplying or dividing the fractions. Indeed, whereas adding (subtracting) two fractions with different denominators requires finding the common denominator followed by their equivalence-aided modification, such fractions do not require any alteration in order to multiply them (in the case of division, the only modification required is the inversion of the divisor). How can one explain this difference? Just as the conceptual meaning of the *Invert and Multiple* rule was explained by demonstrating its meaning in the case of integers (Section 7.6.1), let us consider addition and multiplication of integers in context: *What is the sum of 5 apples and 4 pears*? Numerically, the sum is 9, but contextually, it is neither 9 apples nor 9 pears; instead, we have 9 fruits. That is, the word fruit is the common denomination for apples and pears. Put another way, we have to modify the original words used to contextualize abstract symbols 5 and 4 so that to add fruit and not apples and pears. (Same can be said about subtraction). But in the case of multiplication, we have *repeated* addition and the product 4×5 may not be considered as repeating 5 apples 4-pear times – this just does not make any sense. That is, a contextual denomination may be attached only to the repeated number, 5, seeing the product 4×5 as repeating 5 apples 4 times, something that does make sense as the repetition results in 20 apples. This explains why, procedurally, addition is more complicated than multiplication, although conceptual explanation of an apparently "easy" procedure is not easy (see Sections 7.3-4). Now, what is true for integers, has to be true for fractions. Indeed, one of the most profound characteristic features and big ideas of mathematics is the development of more and more general concepts on higher and higher levels of abstraction, considering the need to protect conceptual coherence of mathematics due to various uses of the concepts in remarkably diverse branches of the discipline [Aleksandrov, 1963]. Teachers of mathematics have to keep this big idea in mind allowing for the context of less abstract concepts (like integers) to be used when explaining the meaning of more abstract ones (like fractions).

7.7 Representing fractions on a number line

Another big idea of mathematics deals with establishing connections between numeric and geometric concepts. To this end, a straight line can be used to represent (real) numbers through the points of the line. Such a straight line each point of which is labeled by one and only one number is called a number line. Reciprocally, any (real) number can be associated with one and only one point on a number line. In other words, one-to-one correspondence between numbers and the points on a number line can be established.

The use of a number line in the elementary school curriculum is connected to the idea of measurement. Although in the United States, the term number line appears first time in grade 3 in connection with fractions, already in grade 1 students learn to "measure lengths ... by iterating length units ... where the object being measured is spanned by a whole number of length units with no gaps or overlaps" [Common Core State Standards, 2010, p. 16]. Therefore, the first step in the process of connecting numbers and points on a straight line is to introduce the unit segment (to be iterated) and define its left and right endpoints as locations for the numbers zero and one, respectively. Then, by iterating this segment to the right along the straight line, the left endpoint of any iterated segment would have the same numeric label as the right endpoint of the adjacent (from the left) segment (Fig. 7.18, bottom). In that way, the location of a whole number on a straight line can be determined by turning the straight line into a (whole) number line. That is, such a number line can be introduced in the first grade to support measurement of segments of different lengths. This activity may motivate young children asking a question about numbers describing the length of segments the right endpoints of which do not coincide with a numeric label of a point on a (whole) number line. In particular, such numbers are fractions.

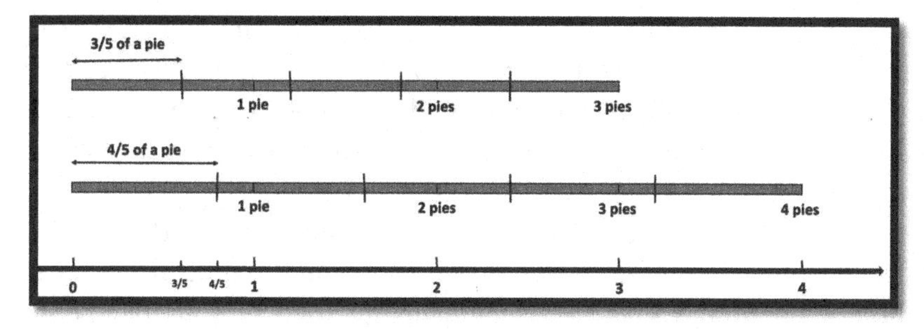

Fig. 7.18. From sharing pies to a number line.

In order to locate fractions on this (originally whole) number line, one has to start with activities that give birth to fractions as numbers. One such activity may be to use the dividend-divisor context for fractions and divide a certain number of (rectangular) pies among a certain number of people. In order to get proper fractions through this process, the number of people should be greater than the number of pies. Consider the case of dividing 3 pies among 5 people fairly. Considering pie as a unit to be iterated, one has to divide three unit segments in five equal pieces. This division is shown in Fig. 7.18 (top) where each segment is measured by the fraction 3/5. The far-left segment is included into the whole pie because when dividing three pies among three people fairly, the resulting piece (a pie) for each person is larger than a piece obtained through dividing three pies among five people. The last statement stems from common sense. In that way, the number 3/5 is located between zero and one.

Likewise, dividing fairly 4 pies among 5 people yields a piece measured by 4/5 of a pie. As shown in Fig. 7.18 (middle), the fraction 4/5 is located between zero and one as well. Because dividing 4 pies among 5 people yields a better deal than dividing 3 pies among 5 people, the fraction 4/5 is closer to the number 1 than the fraction 3/5; that is, the point representing the fraction 4/5 is located between the points representing the fraction 3/5 and the number 1. Similarly, other fractions can be located on a number line, serving as measurements of (many but not all) segments the right endpoints of which do not coincide with a whole number label of a point, although deciding mutual location of different fractions requires a special consideration. From here, as shown in Fig. 7.18 (bottom), a number line as a concept and its use for representing fractions follow.

Using a number line, one can demonstrate the equivalence of fractions. For example, by iterating 5, 10, 15, and 20 times the segment with the fraction 3/5 as the right endpoint one gets to the points representing the numbers 3, 6, 9, and 12, respectively. In other words, dividing 5 into 3, 10 into 6, 15 into 9, and 20 into 12 yield the same number, 3/5. That is, the relations of equivalence, 3/5 = 6/10 = 9/15 = 12/20, hold true. Likewise, by iterating 5, 10, 15, and 20 times the segment with the fraction 4/5 as the right endpoint, one gets to the points representing the numbers 4, 8, 12, and 16, respectively. In other words, dividing 5 into 4, 10 into 8, 15 into 12, and 20 into 16 yield the same number, 4/5. That is, the relations of equivalence, 4/5 = 8/10 = 12/15 = 16/20, hold true.

Furthermore, using a number line provides students with various collateral learning opportunities [Dewey, 1938]. Indeed, this geometric milieu of representing numbers opens a window on many interesting and challenging mathematical problems and motivates one to find ways of using technology to support problem solving. For example, how can one divide a segment in two or three equal parts? Note that if we know how to divide a segment in two equal parts, we can divide it in four, eight, sixteen, and so on equal parts. Likewise, if we know how to divide a segment in two and three equal parts, we can divide it in six, nine, twelve, eighteen, twenty-four, and so on equal pats. Moreover, if we know how to construct a point on the number line representing the fraction measured by $1/n$ of the unit segment, then, by iterating m times a segment with the left and right endpoints zero and $1/n$, respectively, the fraction m/n can be located on the number line as the right endpoint of the last iterated segment.

7.8 Unit fractions as benchmark fractions

In integer arithmetic, the benchmark whole numbers are multiples of ten. A problem may be to position a whole number between two consecutive multiples of ten, e.g., 40 < 47 < 50. In much the same way, unit fractions may be called benchmark fractions and a similar problem is to position a fraction, say, 7/18, between two consecutive unit fractions. One can see that 7/18 < **9**/18 = 1/2 and 7/18 > **6**/18 = 1/3. That is, 1/3 < 7/18 < 1/2 (Fig. 7.19). Alternatively, one can write 7/18 > 7/**21** = 1/3 and 7/18 < 7/**14** = 1/2 . Once again, 1/3 < 7/18 < 1/2 (Fig. 7.20). The inequalities can be interpreted in terms of a fair division of pizzas (or pies). For example, the fractions 7/18 and 9/18 can be interpreted, respectively, as dividing 7 pizzas among 18 people and 9 pizzas among 18 people. It is not difficult to see that the more pizzas we have for the same number of people, the more pizza one can get. Numerically, 7/18 < 9/18 = 1/2. Likewise, the fractions 7/18 and 6/18 can be interpreted, respectively, as dividing 7 pizzas among 18 people and 6 pizzas among 18 people. This time, having fewer pizzas for the same number of people results in a smaller quantity of pizza that one can get. Numerically, we have 7/18 > 6/18 = 1/3. In much the same way, the inequalities 7/18 < 7/14 and 7/18 > 7/21 can be justified within the context of dividing 7 pizzas among 14, 18, and 21 peoples. This is exactly what it means to "bring two complimentary abilities to bear on problems involving quantitative relationships: the ability to *decontextualize* ... and the ability to *contextualize*" [Common Core State Standards, 2010, p. 6, italics in the original].

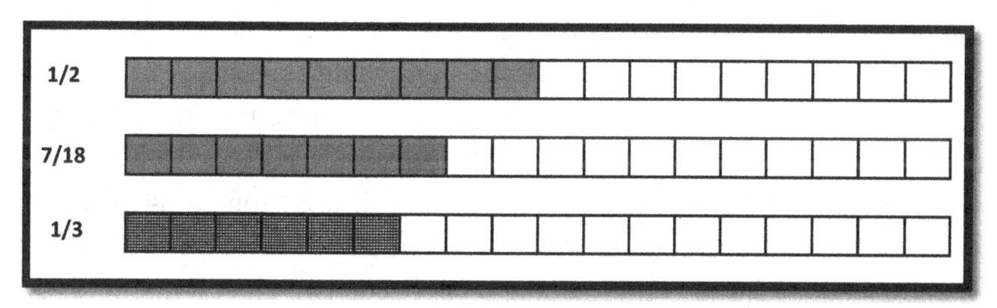

Fig. 7.19. Placing 7/18 between consecutive benchmark fractions.

Fig. 7.20. Alternative placing of 7/18 between 7/14 and 7/21.

Remark 7.4. A recourse to contextualization when placing a common fraction between *consecutive* unit fractions should be carefully examined as context does not always bring about unit fractions that are consecutive benchmark fractions. For example, the relationships $1/5 = 3/15 > 3/16 > 3/18 = 1/6$ can be justified contextually as sharing 3 pizzas with 15, 16, and 18 people, when 15 people get more pizza than 16 people and 18 people get less pizza than 16 people. At the same time, in the case of the fraction 3/16 interpreted as sharing 3 pizzas among 16 people, one can note that sharing 4 and 2 pizzas among 16 people would also result, respectively, in more and less pizza than with three pizzas. Yet, the relationships $1/4 = 4/16 > 3/16 > 2/16 = 1/8$ place 3/16 between the fractions 1/4 and 1/8 which are not *consecutive* unit fractions. Put another way, contextualization and decontextualization go hand by hand requiring continuous validation of their mutual coherence.

It is slightly more difficult to position the fraction 25/84 between two consecutive unit fractions. To do that, one may write $25/84 < 25/75 = 1/3$ and $25/84 > 25/100 = 1/4$ (placing 84 between two closest multiples of 25, $75 = 3 \cdot 25 < 84 < 4 \cdot 25 = 100$, and simplifying fractions to the unit fraction form). Therefore, $1/4 < 25/84 < 1/3$ where 1/4 and 1/3 are two consecutive unit fractions. Sometimes, it is important to calculate the difference between a fraction and the largest benchmark fraction smaller than it. In our case, $25/84 - 1/4 = (25 - 21)/84 = 4/84 = 1/21$ and therefore, $25/84 = 1/4 + 1/21$. The significance of the last equality will be explained in the next section.

Remark 7.5. Note that 1/4 = 1/5 + 1/20 and 1/21 = 1/42 + 1/42. Therefore, 25/84 = 1/5 + 1/20 + 1/21 = 1/5 + 1/20 +1/42 + 1/42. Because any (positive) fraction can be represented as a sum of two fractions in at least one way, such decomposition of a fraction in two fractions may continue as long as one wishes. In other words, any fraction can be represented as a finite sum of other fractions with the number of terms being any given number. This phenomenon is quite different from the case of integers when a positive number n may be partitioned into a sum of like numbers with at most n summands. Furthermore, a fraction may be represented as an infinite sum of other fractions. For example,

$\frac{1}{2}=\frac{1}{4}+\frac{1}{8}+\frac{1}{16}+\frac{1}{32}+\frac{1}{64}+\ldots$ and $\frac{25}{84}=\frac{1}{4}+\frac{1}{25}+\frac{4}{625}+\frac{16}{15265}+\frac{64}{390625}+\ldots$.

One can check to see that the ratio of any two consecutive terms in the infinite sum for 1/2 is equal to 1/2 and in the sum for 25/84 such ratio is equal to 4/25. That is, in both cases we have the sums of infinitely decreasing geometric series with the ratios 1/2 and 4/25, respectively, a topic of the secondary school mathematics curriculum.

Remark 7.6. Another difference between integers and their reciprocals (i.e., unit fractions) is worth noting. Whereas any odd number is a sum of two numbers of different parity, any unit fraction with an odd denominator is a sum of two unit fractions with even denominators. For example, 1/3 = 1/4 + 1/12 = 1/6 + 1/6. This note can be proved using the following simple facts about integers: two integers of the same/different parity have an even/odd sum, the product of two even integers (or of different parity) is even, and the product if two odd integers is odd. The equation $\frac{1}{n}=\frac{1}{a}+\frac{1}{b}$ is equivalent to $ab = n(a + b)$ from where it follows that in the case when n is an odd number, a and b may only be even numbers. Indeed, if at least one of them is odd then both sides of the last equation are numbers of different parity. Thus, both a and b may only be even numbers. At the same time, a unit fraction may not be a sum of two unit fractions with odd denominators.

7.9 Egyptian fractions with applications

History of mathematics can provide interesting mathematical ideas and, according to the Conference Board of the Mathematical Sciences [2012,

p. 61], is worth to "be woven into existing mathematics courses". A history of ancient Egyptian mathematics (around 1650 B.C.) provides such context for both conceptual and procedural mathematical learning. An Egyptian fraction is a sum of the finite number of distinct unit fractions; for example, $\frac{1}{3}+\frac{1}{5}+\frac{1}{7}$ is such a fraction. A unit fraction, like $\frac{1}{3}$, is an Egyptian fraction, although it can be represented as a sum of other unit fractions, e.g., $\frac{1}{3}=\frac{1}{4}+\frac{1}{13}+\frac{1}{156}$. Sometimes, an Egyptian fraction is understood as a special representation of a fraction (alternatively, a positive rational number) through a finite sum of distinct unit fractions, so that the sum $\frac{1}{3}+\frac{1}{5}+\frac{1}{7}$ which is equal to $\frac{71}{105}$ might not considered an Egyptian fraction representation of $\frac{71}{105}$. Indeed, if one asks *Wolfram Alpha* to represent $\frac{71}{105}$ as an Egyptian fraction, the answer is $\frac{1}{2}+\frac{1}{6}+\frac{1}{105}$. However, the term Egyptian fraction can be used for any finite sum of distinct unit fractions.

Apparently, *Wolfram Alpha* uses the so-called Greedy algorithm for Egyptian fractions. This algorithm was first used by Fibonacci (see Chapter 1, Section 1.1, and Chapter 6, Section 6.7) to convert a given non-unit fraction into an Egyptian fraction. The algorithm is based on finding the largest unit fraction smaller than the given fraction, then finding the difference between the two fractions and continue (if the difference is not a unit fraction) the algorithm with the difference. For example, the largest unit fraction smaller than 3/5 is 1/2. Thus $\frac{3}{5}=\frac{1}{2}+x$, whence $x=\frac{3}{5}-\frac{1}{2}=\frac{6}{10}-\frac{5}{10}=\frac{1}{10}$. So, $\frac{3}{5}=\frac{1}{2}+\frac{1}{10}$, where $\frac{1}{2}+\frac{1}{10}$ is an Egyptian fraction representation of the fraction $\frac{3}{5}$. A not Greedy algorithm could be to represent $\frac{3}{5}=\frac{1}{3}+x$, where 1/3 is not the largest unit fraction smaller than 3/5. We have $x=\frac{3}{5}-\frac{1}{3}=\frac{9}{15}-\frac{5}{15}=\frac{4}{15}$. That is, $\frac{3}{5}=\frac{1}{3}+\frac{4}{15}$. In turn, $\frac{4}{15}=\frac{1}{4}+x$ (here we use the Greedy algorithm as 1/4 is the largest unit fraction smaller than 4/15), whence $x = \frac{4}{15} - \frac{1}{4} = \frac{16}{60} - \frac{15}{60} = \frac{1}{60}$. Alternatively, 4/15 = 1/5 + x, whence x = 1/15. That is, 1/3 + 1/4 + 1/60 and 1/3 + 1/5 + 1/15 are another two Egyptian fraction representations of the fraction 3/5. One can also see that Egyptian fraction representation is not unique.

Often, Egyptian fractions can be used to solve simple division word problems more effectively in comparison with the use of the dividend-divisor context (Section 7.1), such as dividing 5 (identical) circular pizzas among 6 people as shown in Fig. 7.21. The equality 5/6 = 1/2 + 1/3, the right-hand side of which is an Egyptian fraction, can be used to decide the division in a different way (Fig. 7.22): 3 pizzas are divided into 2 equal pieces each and 2 pizzas are divided into 3 equal pieces each (rather than each pizza in 6 pieces as in Fig. 7.21). Consequently, each person would get half of a pizza plus one-third of it. The total number of pieces in Fig. 7.21 is 30 and the total number of pieces in Fig. 7.22 is 12.

Remark 7.7. It should be noted that whereas both dividend-divisor context and Egyptian fraction fair division of pizzas provide each person with *identical* pieces (either uniquely identical like in the former case or identical in terms of a number of pieces like in the latter case), it is possible to divide pizzas fairly in terms of the quantity of pizza yet without providing each person with the same number of pieces. Such division may be called a semi-fair division. Fig. 7.23 shows a semi-fair division of 5 pizzas among 6 people by cutting off an identical piece (measuring 1/6 of pizza) from each pizza enabling one person to get 5 such pieces and each of the remaining 5 people to get the rest of each pizza (measuring 5/6 of pizza). One can note that the total number of pieces in which 5 pizzas were divided through this method is 10 – smaller than in the case of dividend-divisor context (30 pieces) and in the case of Egyptian fraction (12 pieces). Having this semi-fair division in mind, consider another example stemming from the equality $\frac{3}{5} = \frac{1}{2} + \frac{1}{10}$ (an Egyptian fraction representation of 3/5) which can be used to decide the division of 3 pizzas among 5 people[23]. The last Egyptian fraction implies that all the pizzas are divided in half and then one of the halves is divided into 5 equal parts so that each person would get one-half of a pizza plus one-tenth of a pizza (the top part of Fig. 7.24). Here, we have 10 pieces of pizza rather than 15 when using the dividend-divisor context for fractions (the middle part of Fig. 7.24). Finally, using semi-fair division (the bottom part of Fig. 7.24), one can divide each pizza into three pieces – $\frac{1}{5}, \frac{1}{5}, \frac{3}{5}$ – thus having 6 pieces measuring 1/5 of pizza and 3 pieces measuring 3/5 of pizza; the total of 9 pieces. Therefore, two people would get 3 pieces measuring 1/5 of pizza and three people would get a single piece measuring 3/5 of pizza. Once again, semi-fair division minimizes the number of pieces.

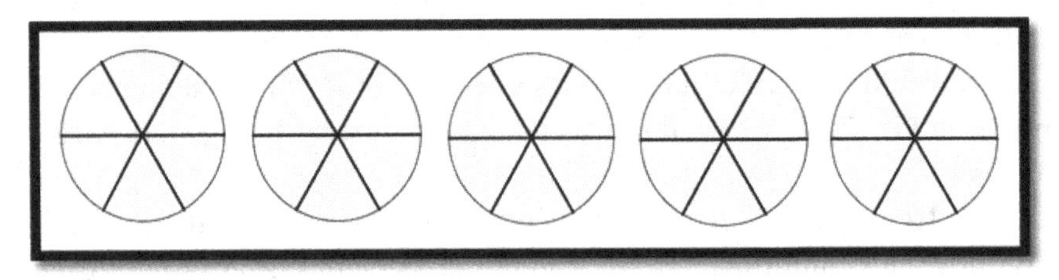

Fig. 7.21. Dividing 5 pizzas among 6 people using dividend-divisor context.

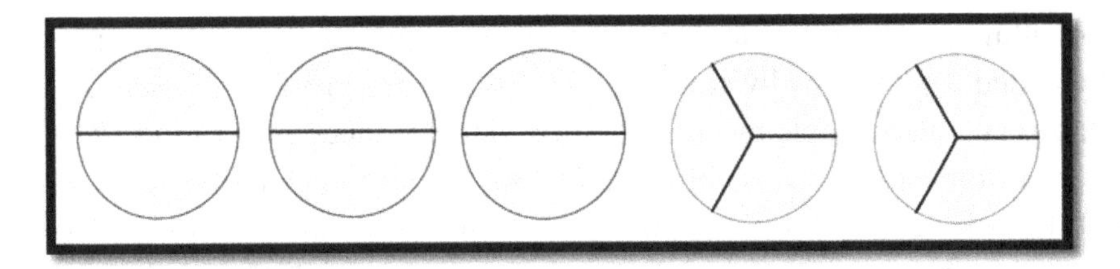

Fig. 7.22. Dividing 5 pizzas among 6 people using the Egyptian fraction $\frac{5}{6}=\frac{1}{2}+\frac{1}{3}$.

---

[23] Note that a distinction between the fractions $\frac{5}{6}$ and $\frac{3}{5}$ is that the former one is a unit fraction short of the whole $\left(1-\frac{5}{6}=\frac{1}{6}\right)$ and the latter one is not $\left(1-\frac{3}{5}=\frac{2}{5}\right)$. This distinction would require an additional insight when accommodating the semi-fair division.

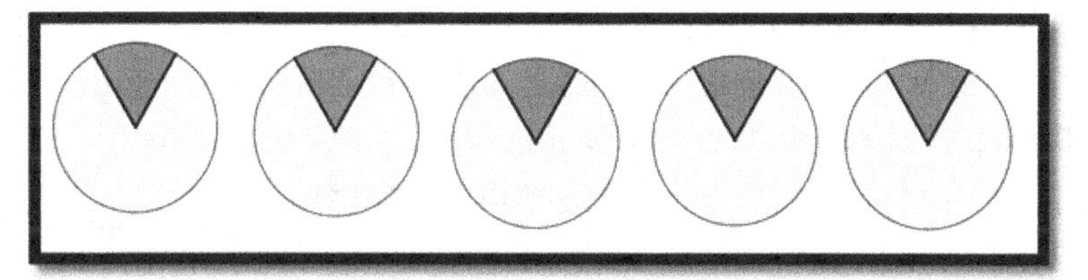

Fig. 7.23. Semi-fair division of 5 pizzas among 6 people:

$$5=\left[\underbrace{\frac{1}{6}+\frac{1}{6}+\frac{1}{6}+\frac{1}{6}+\frac{1}{6}}_{for\ 1\ person}\right]+\underbrace{\frac{5}{6}+\frac{5}{6}+\frac{5}{6}+\frac{5}{6}+\frac{5}{6}}_{for\ 5\ people}.$$

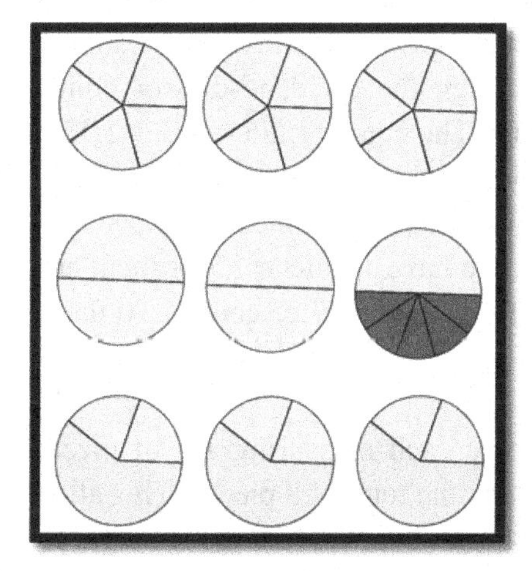

Fig. 7.24. Dividing 3 among 5: dividend-divisor (top), Egyptian (center), semi-fair (bottom).

A more complicated example is provided by the Egyptian fraction representation $\frac{5}{7}=\frac{1}{2}+\frac{1}{5}+\frac{1}{70}$. The dividend-divisor context yields 35 equal pieces by dividing each of the five pizzas into seven equal pieces. The Egyptian fraction yields 21 pieces: 7 pieces each of which is 1/2 of a pizza; then 7 pieces, each of which is 1/5 of a pizza; and another 7 pieces each of which is 1/70 of a pizza. Once again, we have fewer pieces by dividing pizzas through the Egyptian fraction. However, the latter way of dividing pizzas is not intuitively clear in comparison with the first two examples: dividing five pizzas among six people (Fig. 7.22) and three pizzas among five people (Fig. 7.24, center). Note that $7\cdot\left(\frac{1}{2}+\frac{1}{5}+\frac{1}{70}\right)=\frac{7}{2}+\frac{7}{5}+\frac{1}{10}=\frac{175+70+5}{50}=\frac{250}{50}=5$; that is, putting all the 21 pieces together yields 5 pizzas that were divided among seven people in a fair (but rather complicated) way.

Remark 7.8. Comparing the number of pieces resulted from the dividend-divisor context and the Egyptian fraction representation, should not be taken to mean that an Egyptian fraction always provides fewer pieces than the dividend-divisor context. For example, dividing 2 pizzas among 5 people through the dividend-divisor context yields 10 equal pieces (Fig. 7.25 left). The equality 2/5 = 1/3 + 1/15, being the Egyptian fraction representation of the same division, requires dividing each pizza into 3 equal pieces and then divide one such piece into 5 equal pieces. As a result, once again, we have 10 pieces: 5 of them are 1/3 of a pizza and another 5 are 1/15 of a pizza (Fig. 7.25 center). At the same time, the semi-fair division (Fig. 7.25, right) cuts each pizza into 4 pieces – $\frac{1}{5}, \frac{1}{5}, \frac{1}{5}, \frac{2}{5}$ – thus having 6 pieces each measuring 1/5 of pizza and 2 pieces each measuring 2/5 of pizza; the total of 8 pieces. This allows 3 people to get 2 pieces of one type (1/5) and 2 people to get a single piece of another type (2/5). One can see that semi-fair division yields the smallest number of pieces when dividing 2 pizzas among 5 people using three different methods.

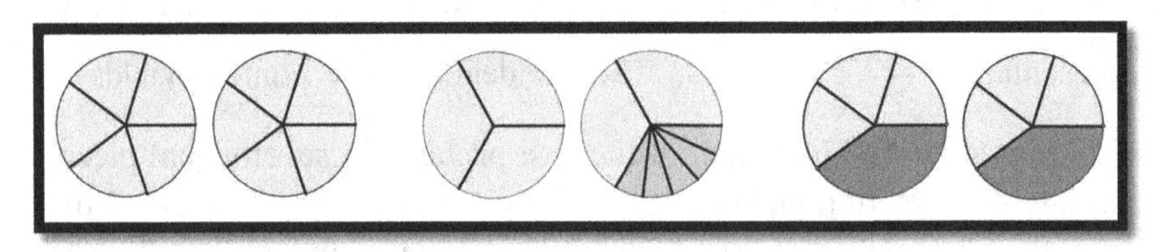

Fig. 7.25. Dividing 2 among 5: dividend-divisor (left), Egyptian (center), semi-fair (right).

Remark 7.9. The number of pizza pieces provided by the Egyptian method can be calculated without actually performing the division. For example, the equality (provided by *Wolfram Alpha*) indicates that each of the seven people would get three (different) pieces of pizza because the right-hand side has *three* addends. Therefore, there will be 21 pizza pieces in all. Likewise, the equality (provided by *Wolfram Alpha*)

$$\frac{7}{43}=\frac{1}{7}+\frac{1}{51}+\frac{1}{3071}+\frac{1}{11785731}+\frac{1}{1852045953417},$$

having *five* unit fractions in its right-hand side (some with very large denominators defying physical division) indicates that each of the 43 people would get 5 different pieces of pizza. Therefore, the Egyptian method of dividing 7 pizzas among 43 people yields 215 (= 43·5) pieces. At the same time, the same type of division of 7 pizzas among 41 people yields 82 pieces due to the equality $\frac{7}{41}=\frac{1}{6}+\frac{1}{246}$.

Remark 7.10. The semi-fair division can be formulated in the following procedural terms. For example, in order to divide 6 identical pizzas among 11 people in a semi-fair way, one has to find the difference between the number of people and the number of pizzas, i.e., 11 – 6 = 5, and then cut off from each pizza 5 pieces each measured 1/11 of pizza, thus having the remaining piece measured 6/11 of pizza, as $1-\left(\frac{1}{11}+\frac{1}{11}+\frac{1}{11}+\frac{1}{11}+\frac{1}{11}\right)=\frac{6}{11}$. In all, we have 30 (small) pieces each measured 1/11 of pizza and 6 (large) pieces each measured 6/11 of pizza. In that way, 11 people can be split in two groups: 5 people in the first group who would get 6 small pieces each and 6 people in the second group who would get a single large piece each. However, the representation 6/11 = 1/2 + 1/22 indicates that dividing 6 pizzas among 11 people using an Egyptian fraction yields 22 pieces, smaller than the semi-fair division (36 pieces). This and other examples considered above show the complexity of ideas associated with a seemingly simple context of dividing identical pizzas using three different methods – the dividend-divisor, the Egyptian fraction, and the semi-fair ones. More information about the interplay between the semi-fair and the Egyptian fraction divisions can be found elsewhere [Abramovich, 2020b].

CHAPTER 8: FROM FRACTIONS TO DECIMALS TO PERCENT

8.1 From long division to fractions as decimals

The algorithm of long division (Chapter 2, Section 2.4) can be used to demonstrate where the decimal representation of a common fraction comes from. The algorithm allows one to complete hidden in the common fraction division and replace it by a number written in a decimal notation the digits of which represent the number of 1/10's, 1/100's, 1/1000's, and so on, included in the fraction. For example, the common fraction 5/16 can be transformed into a decimal fraction by dividing 16 into 5 through long division (Fig. 8.1). The result is the number 0.3125 where the digits 3, 1, 2, and 5 represent, respectively, the number of 1/10's, 1/100's, 1/1000's, and 1/10000's included in 5/16; that is,

$$\frac{5}{16}=0.3125=\frac{3}{10}+\frac{1}{100}+\frac{2}{1000}+\frac{1}{10000}=3\cdot10^{-1}+1\cdot10^{-2}+2\cdot10^{-3}+5\cdot10^{-4}.$$

This representation is similar to representing the integer 3125 as $3 \cdot 10^3 + 1 \cdot 10^2 + 2 \cdot 10^1 + 5 \cdot 10^0$. The right-hand side of Fig. 8.1 also shows the face values 3, 1, 2, and 5 as quotients of four divisions until the zero remainder is reached and thus, if division continues, the remaining quotients would also be zeros. This means that the decimal representation of 5/16 is a terminating decimal. At the same time, 2/5 = 0.4 and, therefore, zero remainder is reached after the first division (of 20 by 5). In particular, one can see that in base-ten arithmetic, not only "non-negative integers represented in base ten can be viewed as 'polynomials in 10'" [Conference Board of the Mathematical Sciences, 2012, p. 59], but common fractions can also be viewed as polynomials in the negative integer powers of ten.

Another conclusion that one can make from observing decimal representations of common fractions is that numeric comparison of the decimals 0.3125 and 0.4 is much easier than that of the fractions 5/16 and 2/5. This is precisely because, while common fractions can be seen representing the dividend-divisor context which extends the division of integers to the domain of fractions when the dividend is not a multiple of the divisor, their decimal equivalents represent notations with completed division. The right-hand side of Fig. 8.1 shows an alternative representation of the long division through the sequence of relations among dividend, divisor, quotient, and remainder (DDQR), in which the quotients are the face values in the decimal representation of the common fraction 5/16.

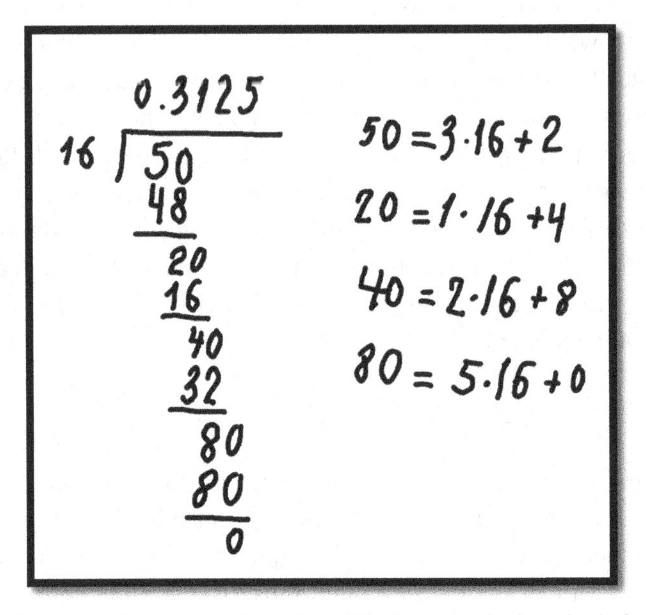

Fig. 8.1. Long division of 5 by 16 with the remainders 2, 4, 8, and 0.

The process of long division can demonstrate that there are fractions for which zero remainder may never be reached. In that case, long division never terminates; however, the behavior of the face values it produces is periodic (i.e., repeating). For example, consider the process of obtaining the decimal representation of the fraction 3/7 through long division. As shown in Fig. 8.2, the long division goes through the sequence of six DDQR relations until the relations begin repeating. That is, this repetition is due to a simple yet conceptually profound fact that when an integer smaller than 7 (e.g., 3) is divided by 7, the corresponding remainder may not be greater than or equal to 7. Indeed, when integers smaller than 7 are divided by 7, only six DDQR relations are possible: $1 = 0 \cdot 7 + 1$, $2 = 0 \cdot 7 + 2$, $3 = 0 \cdot 7 + 3$, $4 = 0 \cdot 7 + 4$, $5 = 0 \cdot 7 + 4$, $6 = 0 \cdot 7 + 6$. Therefore, $3/7 = 0.428571428571428571...$. Such non-terminating periodic (repeating) decimal has a special notation: $3/7 = 0.\overline{428571}$ with the repeating part of the decimal having a bar above it. Likewise,

$1/3 = 0.3333... = 0.\overline{3}$, $2/3 = 0.6666... = 0.\overline{6}$, $1/11 = 0.090909... = 0.\overline{09}$.

In general, when an integer smaller than n is divided by n, the *largest* number of different remainders is equal to $n-1$. The corresponding quotients become face values and they repeat each other by forming cycles the length of which may not be greater than $n - 1$. In the case when the prime factorization of n consists of the powers of two and five only, the common fraction with the denominator n is represented by a terminating decimal; otherwise, the fraction is represented by a non-terminating decimal the face values of which form cycles (repeating strings of digits) of length not greater than $n-1$. In other words, through the process of long division by 7, one can see that the process is repeating same calculations over and over. Furthermore, one can recognize that the corresponding fraction is represented by a repeating non-terminating decimal. Through developing such representation, one can understand what it means to "express regularity in repeated reasoning" [Common Core State Standards, 2010, p. 8].

One can also use *Wolfram Alpha* to see that the periods (lengths) of cycles in the decimal representations of the unit fractions with prime number[24] denominators, say, 1/13, 1/23, 1/31, 1/37, and 1/41 are, respectively, 6, 22, 15, 3, and 5. One may note that each of the five denominators is one greater than a multiple of the corresponding period. In other words, dividing each denominator by the corresponding period of the repeating string of digits yields the remainder one. However, even recognizing this seemingly opaque pattern does not allow one to guess that the number 16 is the period for the fraction 1/17. This demonstrates the complexity of mathematics behind the issue of relating the length of a cycle in the decimal representation of $1/n$ to n; something, that is beyond the scope of this textbook.

---

[24] A prime number is an integer with *exactly* two different divisors, one and itself. Elementary teacher candidates often confuse odd numbers and prime numbers. For example, 9 is an odd number but not a prime number as it has three divisors: 1, 3, and 9.

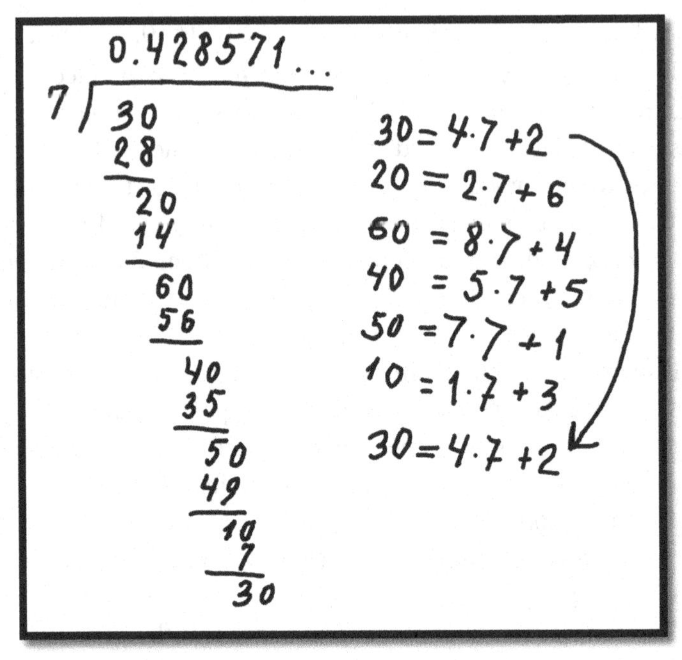

Fig. 8.2. Long division by 7 with 6 different (non-zero) remainders.

8.2 From alternative comparison of fractions to percent

A rational number, two representations of which are fraction (including Egyptian fraction) and decimal, has another representation studied at the elementary level. This third representation is called percent. The term is well known to elementary school students from a number of sources. Many students even refer to certain conditions in terms of percent when talking about weather; e.g., fifty percent chance of snow (or rain). But just as the concept of chance (Chapter 12), the concept of percent has to be formally introduced before it can become an alternative representation of a fraction. One way to introduce percent is through the comparison of fractions using a 100-cell grid. However, the effectiveness of comparing fractions on a 100-cell grid depends on their denominators, even if they are small integers. As an example, consider the fractions 3/5 and 7/10 – which one is bigger? To answer this question, both fractions can be represented as parts of a 100-cell grid shaped in the form of square. Because 1/5 and 1/10 of such grid cover, respectively, 20 and 10 cells, the fraction 3/5 covers 60 cells and the fraction 7/10 covers 70 cells (Fig. 8.3). Therefore, $7/10 > 3/5$.

Consequently, one can say that by representing 3/5 on a 100-cell grid (Fig. 8.3, left), 60 cells per 100 cells were shaded; or, using a Latin word *centum* (*hundred* in English), 60 percent of a 100-cell grid are shaded. The word percent has the following (well-known) symbolic notation: %. Therefore, 3/5 = 60%. Likewise, 7/10 = 70%. Alternatively, 3/5 = 60/100 = 60% and 7/10 = 70/100 = 70%.

Note that the denominators 5 and 10 can easily be multiplied to become 100; indeed, multiplying 5 by 20 yields 100 and multiplying 10 by 10 yields 100 as well. In other words, the denominators of the fractions 3/5 and 7/10 are integer factors of 100. Denominators may not necessarily be integer factors of 100. For example, $100 = 2.5 \cdot 40 = 1.25 \cdot 80$ and, therefore,

$$3/40 = 3 \cdot 2.5/(2.5 \cdot 40) = 7.5/100 = 7.5\%$$

and

$$7/80 = 7 \cdot 1.25/(1.25 \cdot 80) = 8.75/100 = 8.75\%.$$

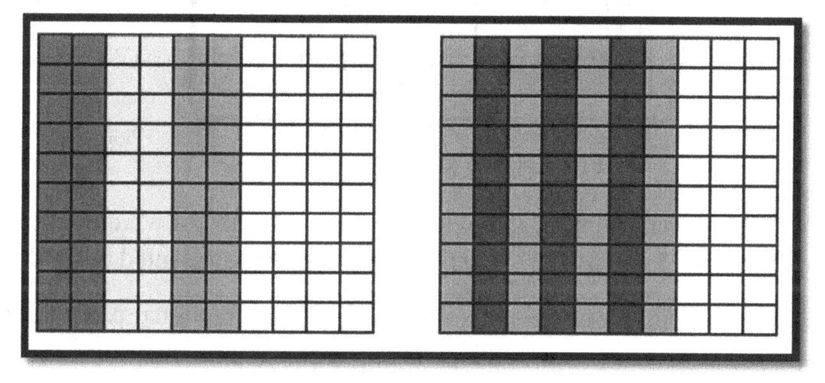

Fig. 8.3. 3/5 = 60% and 7/10 = 70%.

Certain fractions, the denominators of which are not multiples of 2 and/or 5 can still be easily represented on a 100-cell grid. Consider the fraction 1/3. One can see (Fig. 8.4) that 33 cells represent approximately 1/3 of a 100-cell grid with another 66 cells representing approximately 2/3 of the grid. The remaining cell, which is 1% (of the grid), can be divided into three equal parts, each of which is 1/3% (of the grid). Therefore,

$\frac{1}{3} = 33\frac{1}{3}\% = 33.\overline{3}\%$ and $\frac{2}{3} = 66\frac{2}{3}\% = 66.\overline{6}\%$. One can see that

$$\frac{1}{3} = \frac{100}{3}\% = 33\frac{1}{3}\% \text{ and } \frac{2}{3} = 2 \cdot \frac{1}{3} = 2 \cdot \frac{100}{3}\% = \frac{200}{3}\% = 66\frac{2}{3}\%.$$

That is, $\frac{a}{b} = a \cdot \frac{1}{b} = a \cdot \frac{100}{b}\% = \frac{100 \cdot a}{b}\%$.

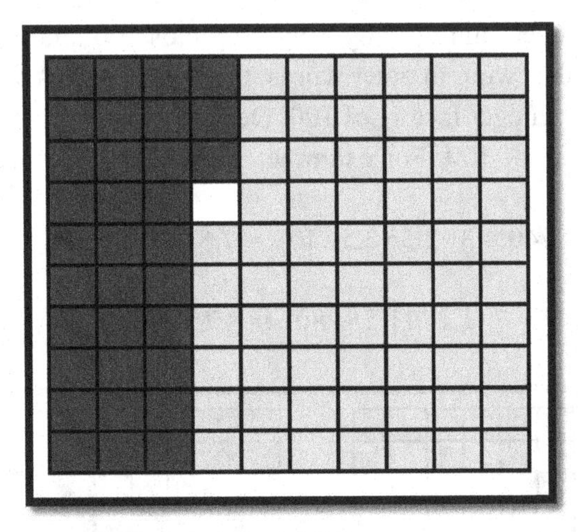

Fig. 8.4. 1/3 and 2/3 as percent.

However, unlike 1/3, the decimal representation of which is $0.\overline{3}$, the fraction 1/7 = $0.\overline{142857}$ cannot be effectively represented on a 100-cell grid. Indeed, because $\frac{1}{7} = \frac{100}{7}\% = 14\frac{2}{7}\%$, the fractional part of the last (percentage) number, i.e., $\frac{2}{7}\%$, does not have a visually lucid representation on a 100-cell grid in comparison with $\frac{1}{3}\%$ (which is just 1/3 of a cell). This, however, does not mean that percentage representation of the fraction 1/7 is difficult to obtain: keeping in mind that 1 = 100%, the difference is in the notation used as the unit, 1 vs. 100%, the fractions of which have to be computed. To conclude this section, note that common fractions may have two different decimal representations. For example,

$\frac{1}{2} = 0.5$. Also, $\frac{1}{2} = 0.4999\ldots = 0.4\overline{9}$. Indeed, due to a well-known (from the secondary mathematics curriculum) formula for the summation of infinite geometric series (see Chapter 6, Section 6.3), we have $0.0\overline{9} = \frac{9}{10^2} + \frac{9}{10^3} + \frac{9}{10^4} + \cdots = \frac{9/10^2}{1-1/10} = \frac{9}{100-10} = \frac{9}{90} = \frac{1}{10} = 0.1$, whence $0.4\overline{9} = 0.4 + 0.0\overline{9} = 0.4 + 0.1 = 0.5 = \frac{1}{2}$. Likewise, $\frac{1}{5} = 0.2$ and $\frac{1}{5} = 0.1\overline{9}$ as $0.0\overline{9} = 0.1$, whence $0.1\overline{9} = 0.1 + 0.1 = 0.2 = \frac{1}{5}$.

8.3 Word problems with fractions, decimals, and percent

In this section, a few word problems involving non-integer quantities will be discussed. Visual strategies will be used in solving such problems with friendly numbers for which a picture can be drawn. It will be shown that in order to use a picture correctly as a thinking device, one must possess conceptual understanding of a mathematical situation involved. Sometimes, in the absence of conceptual understanding, the use of a picture can lead to a misconception. To begin, consider

Problem 8.1. *Donna has $8\frac{1}{3}$ pounds of flour. It takes $\frac{5}{6}$ pounds to bake a cake. How many cakes can Donna bake?*

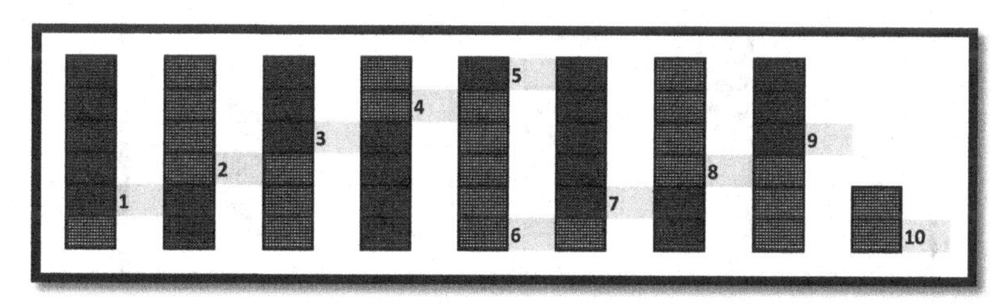

Fig. 8.5. Measuring a fraction by a fraction yields an integer.

Solution. The diagram shown in Fig. 8.5 represents a pictorial solution, where each tower represents a pound which serves as the unit. The far-right tower represents 1/3 of the unit. Dividing each large tower into six equal parts and the smaller tower in two equal parts makes it possible to measure all flour by a quantity required for a cake. In the diagram, all flour is represented by 50 pieces which are measured by 5 pieces needed for a cake. This measurement process yields 10 cakes. In order for a pictorial solution to inform a numeric solution with arbitrary numbers, one has to describe Fig. 8.5 through a numeric equality involving the problem data and the resulting answer. Seeing the diagram of Fig. 8.5 as measurement model for division yields the equality $8\frac{1}{3} \div \frac{5}{6} = 10$ in which the right-hand side was originally unknown. Seeing the diagram of Fig. 8.5 as repeated addition yields the equality $10 \cdot \frac{5}{6} = 8\frac{1}{3}$ in which the first factor of its left-hand side was originally unknown. In the case of using the latter equality for solving the problem, one uses an equation with a missing factor which can be found by dividing the known factor into the known product.

Problem 8.2. *Anna has* $4\frac{2}{3}$ *identical bottles of milk. A glass of milk measured* $\frac{5}{6}$ *of a bottle is a serving. How many servings of milk can Anna make?*

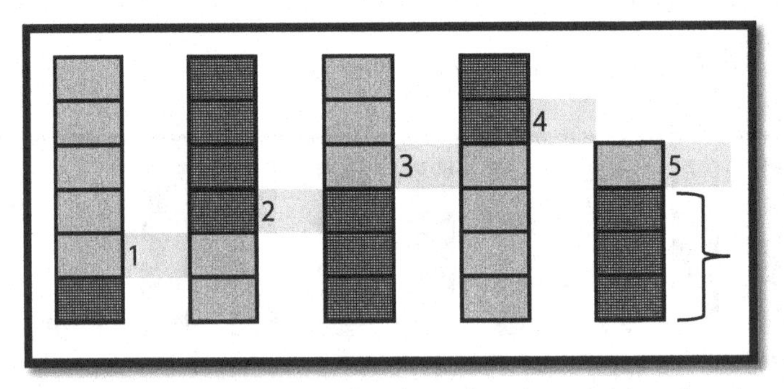

Fig. 8.6. Measuring a fraction by a fraction yields a fraction.

Solution. The diagram shown in Fig. 8.6 represents a pictorial solution, where each large tower stands for a bottle which is the unit in the mixed fraction expressing the quantity of bottles. The smaller tower represents 2/3 of this unit. Dividing each unit into six equal parts turns each unit into 6/6 = 1. Therefore, the smaller tower has to be represented as a combination of 1/6's; that is, the smaller tower can be expressed as 4/6 = 2/3. In the diagram of Fig. 8.6, all milk is represented by 28 equal pieces which have to be measured by 5 such pieces. The process of measurement leads to 5 full glasses and 1/2 of a bottle. Here is where a misconception caused by the use of visual thinking in the absence of conceptual understanding frequently occurs. Is the remaining fractional part of the bottle 1/2 or not (to be added to 5 in order to have another mixed fraction as the answer)? Because the answer has to be given in terms of servings and the number 5 resulting from measuring 28 by 5 has a glass, not a bottle, as the unit, the remaining part of the bottle, namely 1/2, has to be represented as a fraction of this unit.

Therefore, out of $4\frac{2}{3}$ bottles of milk, Anna can make $5\frac{3}{5}$ servings. Note that in Problem 8.1, flour was measured by cakes and the result, an integer, was expressed through cakes. Likewise, measuring milk by glasses (servings) yields a fraction expressed through servings. At the same time, a formal numeric solution $4\frac{2}{3} \div \frac{5}{6} = \frac{14}{3} \cdot \frac{6}{5} = \frac{28}{5} = 5\frac{3}{5}$ does not result in any misconception regarding which unit is involved in the final answer.

Problem 8.3. *Alex paid for a new computer $1,400. It was on sale for 12.5%. How much money did Alex save through this discount?*

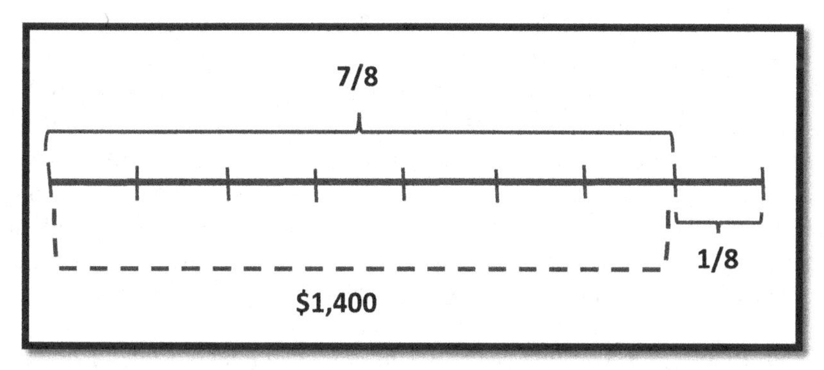

Fig. 8.7. A pictorial solution made possible by friendly numbers.

Solution. This problem also has friendly numbers and can be solved using a picture. However, before drawing a picture, the percentage number has to be converted into a common fraction which, in turn, would inform the drawing. Formally, we have 12.5% = 12.5/100 = 1/8. At the same time, one can recognize in 12.5% a half of 25% and the latter is commonly known as the fraction 1/4. Consequently, half of 1/4 is 1/8. In order to show the fraction 1/8 on a picture, one can draw a segment divided into 8 equal sections, one of which is 1/8 of the segment (Fig. 8.7). Therefore, the discounted price of $1,400 is 7/8 of the pre-sale price. Now, one has to partition $1,400 into seven equals parts each of which is 1/8. This partition yields $200 as 1/8 of the entire segment as well as the amount of money Alex saved.

A formal solution of this problem to allow for the use of arbitrary numbers would take several steps. First, one has to convert the percentage number into a common fraction: 12.5% = 12.5/100 = 1/8 (in our case, we have a unit fraction). The second step is to find the fraction of the unknown pre-sale price x that was paid for the computer: 1–1/8 = 7/8. The third step is to construct the equation connecting two prices, the pre-sale price and the discounted price: $\frac{7}{8}x = 1400$. The fourth step is to solve this equation as an equation with a missing factor:

$$x = 1400 \div \frac{7}{8} = 1400 \cdot \frac{8}{7} = 200 \cdot 8 = 1600.$$

The fifth (final) step is to subtract the discounted price from the pre-sale price as a way of answering the question posed by the problem: 1600 – 1400 = 200.

One can see that Problems 8.1, 8.2, and 8.3 all have two numbers in their formulation; however, whereas the first two problems can be solved through a single step, the last problem required five steps to get the answer. The importance of the last comment deals with a common misconception about solving word problems: one only has to remember a single operation to be applied to the given numbers in order to get the answer. More often than one might think, this is not the case and problem solvers must be aware of the importance of not approaching each problem by simply being curious into which formula to plug given numbers.

Problem 8.4. *A university has to pave two parking lots, A and B, shaped as squares of sides 60 meters and 70 meters, respectively. It is* *known that* $83\frac{1}{3}\%$ *of lot A and* $57\frac{1}{7}\%$ *of lot B need new pavement. Which parking lot needs more pavement?*

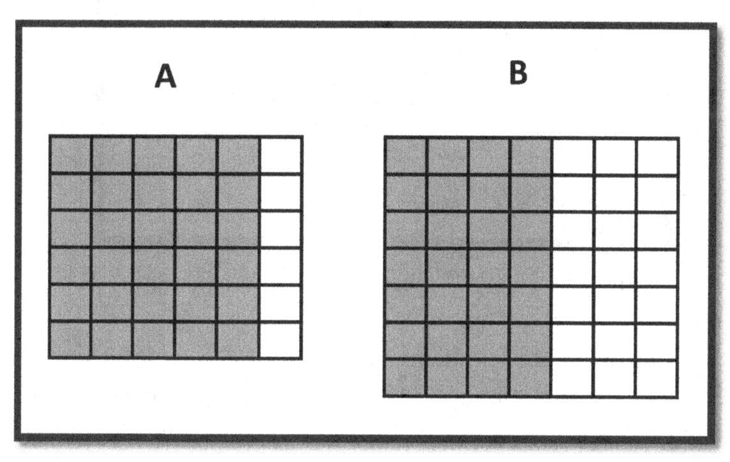

Fig. 8.8. Parking lot A needs more pavement.

Solution. Sometimes, percentage numbers don't look friendly in order to draw a picture unless one takes an effort to reveal their hidden nature in terms of fractions. Indeed, the following two conversions of percentages into common fractions demonstrate just that.

$$83\frac{1}{3}\% = \frac{83+\frac{1}{3}}{100} = \frac{250}{300} = \frac{5}{6}, \quad 57\frac{1}{7}\% = \frac{57+\frac{1}{7}}{100} = \frac{400}{700} = \frac{4}{7}.$$

Now, one can draw two square-shaped grids (Fig. 8.8) with sides measured by six and seven units (each units representing 10 meters) and shade, respectively, 5/6 and 4/7 of the grids. As shown in Fig. 8.8, parking lot A needs 30 square units of pavement and parking lot B needs 28 square units of pavement, each unit measured by 100 m^2. That is, lot A needs move pavement.

Problem 8.5. *A university has to pave two parking lots, A and B, shaped as squares of sides 88.2 meters and 86.1 meters, respectively. It is known that 77.2% of lot A and 80.8% of lot B need new pavement. Which parking lot needs more pavement?*

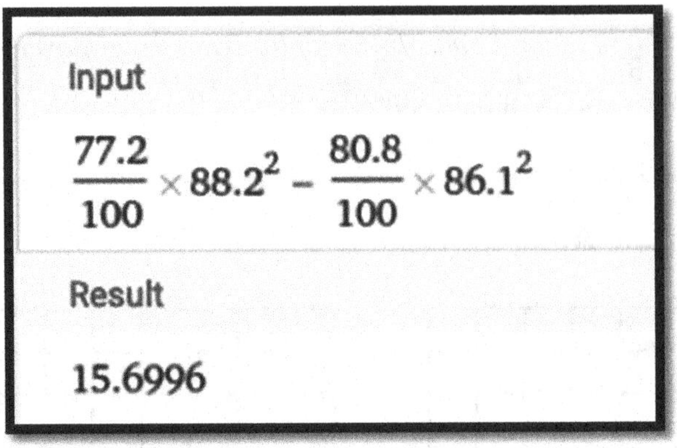

Input

$$\frac{77.2}{100} \times 88.2^2 - \frac{80.8}{100} \times 86.1^2$$

Result

15.6996

Fig. 8.9. Using *Wolfram Alpha* as a calculator.

Solution. In this problem, the sides of the parking lots are not represented by friendly numbers in order to allow for solving the problem by drawing a picture. Instead, one can use *Wolfram Alpha* as a calculator. The result is shown in Fig. 8.9 – parking lot A needs more pavement by about 16 m^2.

CHAPTER 9: RATIO AS A TOOL OF COMPARISON OF TWO QUANTITIES

9.1 Different definitions of ratio

What is ratio? Often the ratio of two numbers is defined as a characteristic which shows how many times one number contains another number; in other words, a ratio is defined as a result of measuring one quantity by another quantity. For example, one can measure one side of a rectangle by another (adjacent) side of the rectangle. In the case when the larger side length of a rectangle is twice as large as its smaller side length, by using the latter to measure the former one can say that the ratio of the side lengths is two to one. As ratio is a number, in that case the ratio is equal to the number 2. One can also measure the smaller side length by the larger side length and say that their ratio is one to two. Just as in the case of using measurement context to introduce a fraction, the ratio in the latter case is the number 1/2. That is, as a common fraction, ratio is the result of dividing one number by another number when the larger number is considered as the whole and the smaller number is considered as its part. One can recognize the part-whole context for a ratio here because the whole is divided in two equal parts, one of which is 1/2 of the whole. According to Common Core State Standards [2010, p. 39], students in Grade 6 "expand the scope of problems for which they can use multiplication and division to solve problems, and they connect ratios and fractions". A classic example of a ratio is the so-called Golden Ratio, $\frac{1+\sqrt{5}}{2}$, when, for example, in a regular pentagon (Fig. 9.1) one measures its diagonal (AC) by its side (BC) and concludes (although in abstract form) that the diagonal, being longer than the side, includes the latter $\frac{1+\sqrt{5}}{2} \approx 1.61803$ times. This is another example of invariance as a big idea of mathematics (see Chapter 1, Section 1.1, and Chapter 10, Section 10.5.1) – regardless of the size of a regular pentagon, the ratio of its diagonal to its side is the Golden Ratio. Using a dragging feature of dynamic geometry software, one can nicely, yet informally, demonstrate this property of a regular pentagon by altering its size and orientation. A formal demonstration of this invariance is beyond the scope of the textbook and can be found elsewhere (Abramovich, 2010, p. 123).

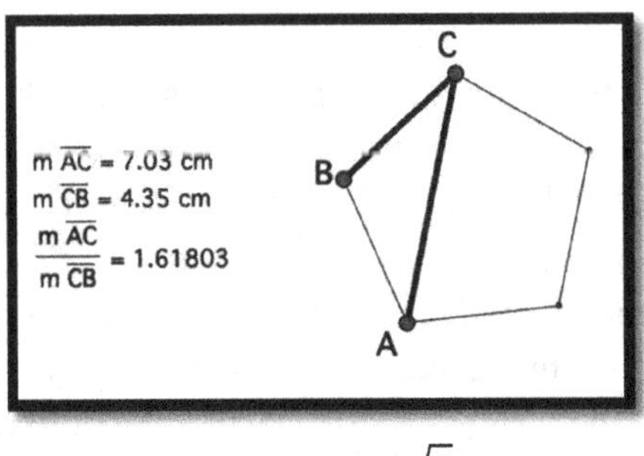

Fig. 9.1. AC/BC $= \frac{1+\sqrt{5}}{2} \approx 1.61803$.

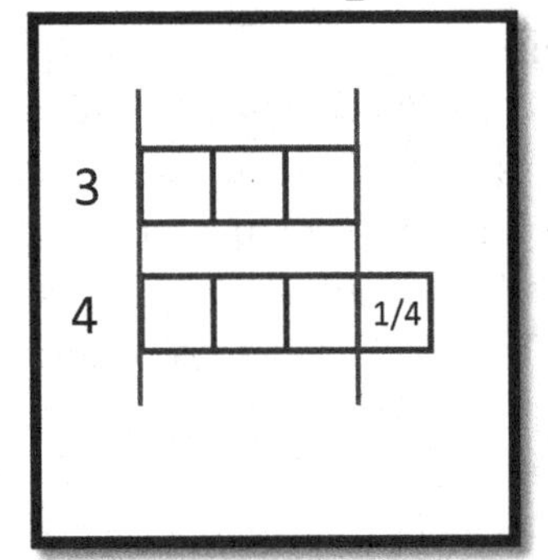

Fig. 9.2. Measuring 3 by 4 yields 3/4 as the ratio 3 to 4.

Understanding of how to measure one number by another number (i.e., what it means numerically for one number to include another number) depends on an interpretation of this measurement (inclusion). For example, one can interpret the process of measuring the number 3 by the number 4 as follows. Let us measure a pie cut into three pieces by a larger pie cut into four pieces, assuming that all pieces are identical as shown in Fig. 9.2. When 3 is measured by 4, this operation can be put in a pictorial context from which it follows that 3 is 3/4 of 4. That is, 4 as a measuring tool is comprised of four equals parts, and 3 as an object of measurement is comprised of three such parts. One of the parts that form 4 is 1/4 of 4; therefore, measuring 3 by 4 yields numerically 3/4 of 4. In other words, the ratio of 3 to 4 is 3/4 as the result of measuring 3 by 4. Likewise, measuring 5 by 7 yields the ratio of 5 to 7, which is equal to 5/7; measuring 6 by 11 yields the ratio of 6 to 11, which is equal to 6/11; and so on.

Division of two numbers can be introduced in the partition context also. So is the ratio and its interpretation. For example, it is often said that the ratio of students to computers in a school is, say, two to one. This statement can be interpreted as the case when students are "partitioned" among computers; for instance, 10 students can share 5 computers evenly yielding two students for a computer (just as 10 apples can be partitioned among 5 people yielding 2 apples for a person). The same numeric outcome (i.e., the ratio of students to computers) would be in the case of 100 students and 50 computers, or 200 students and 100 computers. In the reciprocal relation, the ratio of computers to students in all mentioned cases is one to two; that is, the reciprocal ratio is equal to the fraction 1/2 (alternatively, 0.5 or 50%).

9.2 Introducing ratio as a tool

Whatever a context, ratio can be introduced as a tool of comparing two quantities. This tool is the quotient which, when expressed as a fraction, is written in the simplest form or when the latter form is described in terms of the dividend-divisor context for fractions (Chapter 7, Section 7.1). That is, the ratio 10 to 3 is the fraction 10/3 understood as the number 3 divided into the number 10, or contextually, 10 identical, physically divisible objects (e.g., pizzas or cakes) divided among 3 people fairly (that is, measuring 10 by 3). Likewise, the ratio 3 to 10 is understood as the fraction 3/10 which, in turn, can be interpreted as the number 10 divided into the number 3, or, contextually, 3 identical, physically divisible objects divided into 10 equal parts (that is, partitioning 3 among 10). Obviously, 3 computers cannot be divided into 10 parts to allow each student to have 3/10 of a computer to use. But the ratio shows a possible application of the dividend-divisor context for fractions when two quantities have to be compared.

One may recall that the difference may be used as a tool to compare quantities (Chapter 2, Section 2.1). For example, the pairs of numbers (5, 2), (6, 3) and (10, 7) are in the same difference relation. Indeed, $5 - 2 = 6 - 3 = 10 - 7 = 3$. A context for using difference as a tool of comparison may be as follows. When among 5, 6 and 10 apples there are, respectively, 2, 3 and 7 red apples, each collection of apples has the same number of apples that are not red. And here 2 is a part of 5, 3 is a part of 6, and 7 is a part of 10. But there may be other pairs of numbers, like (5, 2), (10, 4) and (15, 6), which form equal fractions: 5/2, 10/4 and 15/6 (or 2/5, 4/10 and 6/15). Yet, contexts for those pairs are different and ratio can be used for comparing quantities when 2 is *not* a part of 5, 4 is *not* a part of 10 and 6 is *not* a part of 15. Indeed, the fraction 5/2 when interpreted as the result of dividing 5 by 2 (or 2 divided into 5) can describe baskets with 5 oranges and 2 apples, 10 oranges and 4 apples, 15 oranges and 6 apples, or 50 oranges and 20 apples. Whereas in each basket apples are *not* a part of oranges, all such baskets are the same in terms of having oranges to apples ratio equal to the fraction 5/2 or the decimal 2.5 or the percent 250%. Likewise, the reciprocal ratio can be written as 2/5, 0.4, or 40%.

9.3 Using ratio to find an unknown quantity

Consider

Problem 9.1. *The ratio of dogs to cats in a household is 1 to 2. If there are three dogs in the household, how many cats are there*?

To solve this problem, one can begin with the smallest household in which the ratio of dogs to cats is 1 to 2. Fig. 9.3 shows one dog and two cats. If one measures the frame which includes a dog with the frames which include cats, the result of this measurement is 1/2. When there are three dogs in a household (Fig. 9.4), then, in order to preserve the 1 to 2 ratio each new dog would bring two new cats; thus, for three dogs (a new unit) there are six cats (two new units).

Fig. 9.3. The smallest household with the dog to cat ratio equal 1 to 2.

Fig. 9.4. Each dog brings two cats to preserve the 1 to 2 ratio.

On a formal level, there is an unknown number of cats, x, and the known number of dogs, 3. The ratio of 3 dogs to x cats is the fraction $3/x$. But this ratio is equal to $1/2$. The two ratios may be equated in the form called proportion, $\frac{3}{x} = \frac{1}{2}$, an equation between two ratios. To find x from this equation, both ratios (i.e., fractions) can be written with the same denominators, $\frac{3 \cdot 2}{x \cdot 2} = \frac{1 \cdot x}{2 \cdot x}$, whence $x = 3 \cdot 2 = 6$. Alternatively, three dogs can be presented in the form of a single unit, that is, three dogs for a unit. Consequently, x cats have to be presented in the form of two units, $x/2$ cats for a unit. This results in the proportion $\frac{x}{2} = \frac{3}{1}$ whence, once again, $x = 3 \cdot 2 = 6$.

Problem 9.2. *The ratio of dogs to cats in a shelter is 3 to 4. If there are 15 dogs in the shelter, how many cats are there*?

Like in the previous problem, we begin solving the problem with the smallest shelter within which the ratio of dogs to cats is 3 to 4. As shown in Fig. 9.5, fifteen dogs can be presented in the form of three units, five dogs in each unit. To preserve the ratio 3 to 4, there has to be four units with the total unknown number, x, of cats. That is, $\frac{15}{x}=\frac{3}{4}$ whence $x=\frac{15}{3}\cdot 4=20$. Here, the fraction 15/3 shows that fifteen dogs were put in three groups (units) to create a new unit of five dogs. Therefore, one must have four units of cats with five cats in a unit.

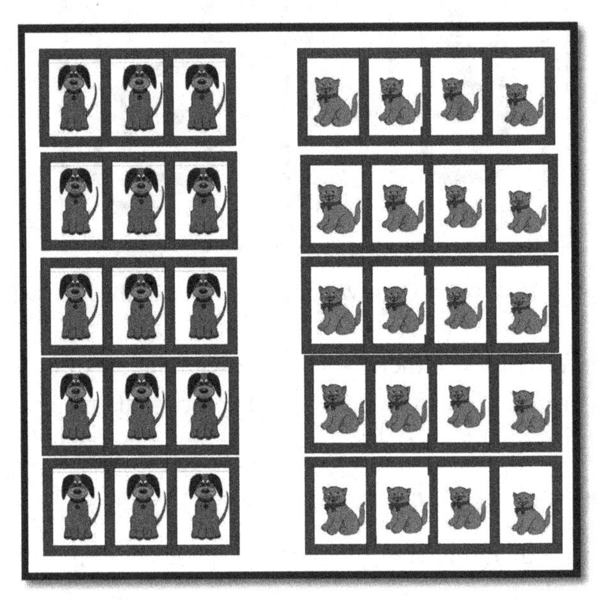

Fig. 9.5. Each triple of dogs brings quadruple of cats.

Problem 9.3. *The ratio of students to professors in a community college is 51 to 4. If there are 848 professors, how many students are there?*

Obviously, one cannot support solving this problem by drawing a picture. We did use picture in the case of small (and arithmetically friendly) numbers and, in doing so, attempted to develop conceptual understanding of a formal problem-solving strategy. In the case of large numbers, an algorithm of constructing a proportion between two ratios has to be used as an applicable strategy. If x is an unknown number of students in the college, then the proportion $\frac{x}{848}=\frac{51}{4}$ holds, whence $x=\frac{51}{4}\cdot 848=10{,}812$.

Remark 9.1. One may note that had we have, instead of (arithmetically friendly) number 848, the number 849 (which is not arithmetically friendly by not being a multiple of 4), the number of students would not be an integer. Likewise, in the case of having 16 dogs (Problem 9.2), the number of cats would not be an integer. Obviously, not all real-life problems have arithmetically friendly data. This points at the importance of formulating a real-life problem to include data which is not necessarily arithmetically friendly. How would one formulate the community college problem, if the number of professors is not divisible by 4? In that case, using the word "about" (or "approximately") in front of a ratio may help. Saying that the ratio of the unknown number of students to 849 professors is about 51 to 4 would yield $10{,}824.75 \cong 10{,}825$. Indeed, when the number of students is 10,825 and the number of professors is 849, their ratio is equal to $10{,}825 \div 849 \cong 12.7503 \cong 12.75 = 51 \div 4$.

9.4 Problems that require insight to avoid an error in using proportional reasoning

Fig. 9.6 shows a three-step high ladder in which each step requires three bars. One can say that the ratio of steps to bars is 1 to 3. The following question may be formulated: *How many bars are needed for a 25-step high ladder?* To answer this question, the proportion $\frac{3}{1}=\frac{x}{25}$ can be constructed, whence $x = 75$.

One can see that the number of bars (b) and the number of steps (s) are in the (proportional) relation $b = 3 \cdot s$. That is, the number 3 is the bars to steps ratio. In general, it can be said that the variables x and y are in the same ratio if there exists number n such that $y = n \cdot x$. As a common extension of the ladder problem, consider the ladder shown in Fig. 9.7. The question to be answered is: *How many bars are needed for the 25-step high ladder?* If one already knows that the 25-step ladder extension shown in Fig. 9.6 is 75, then one has to add just one bar to 75 bars to have 76 as the answer. But this is an insightful solution. Sometimes, the ratio approach is applied to the ladder of Fig. 9.7 noting that the ratio of bars, x, to steps, 25, is 4 to 10. So, by constructing the proportion $\frac{10}{4}=\frac{x}{25}$, one ends up with $x = 62.5$. But x has to be a whole number, unlike 62.5. An error is conceptual, not computational. It is due to the fact that the number of steps and the number of bars satisfy the relation $b = 3 \cdot s + 1$ and, therefore, proportional reasoning may not be applied to the ladder of Fig. 9.7, unless one solves the problem without the bottom bar and then adds 1 to 75. The error is another example of a misconception caused by the lack of conceptual understanding when a proportion may be formed: two variable quantities are in direct proportional relationship when their ratio is a constant. For example, a diagonal and a side of a regular pentagon are in a direct proportional relationship as their ratio is the Golden Ratio. As an aside note that when the product of two variables is a constant, the variables are inversely proportional to each other. For example, given area of a rectangle, its length and width form inverse proportional relationship – the larger the length the smaller the width (see Chapter 10, Section 10.7, Fig. 10.21).

As mentioned in the Standards for Preparing Teachers of Mathematics in the United States [Association of Mathematics Teacher Educators, 2017, p. 86], "beginning teachers ... [should] be aware of likely misconceptions, and be open to understanding unique ways that students might use to express characteristics and generalizations". Likewise, the Conference Board of the Mathematical Sciences [2012, p. 2] recommends that teacher candidates "need the ability to find flows in students' arguments and to help their students understand the nature of errors". Mathematics educators in South Africa emphasize the need for teachers to be able to provide "full and proper explanations ... so that the procedures can be properly generalized and established in the minds of learners" [Department of Basic Education, 2018, p. 39]. In Japan, teachers are encouraged to "help students appreciate the value of using proportional relationships to solve problems efficiently and foster their attitude to willingly use proportional relationships when solving everyday problems" [Takahashi et al., 2004, p. 311]. For more information on the erroneous use of proportional reasoning see [Abramovich and Brouwer, 2011].

Fig. 9.6. Ladder without the bottom rod.

Fig. 9.7. Ladder with the bottom rod.

9.5 Rate as a special ratio

9.5.1 Different problems involving rate

There is a class of the so-called work problems. In such problems two quantities are given: the amount of work one has to do and time one needs to complete this work. For example, it is known that a baker spent four hours to make 100 cakes. The question is: how can one measure the baker's capacity of making cakes? To answer this question, one can find the ratio of the number of cakes to the number of hours spent on making the cakes; that is, to divide 100 by 4 to get 25. In other words, the ratio 100/4 = 25, where the number 25 is the ratio of 100 cakes to 4 hours implying that the baker's capacity can be described as 25 cakes per 1 hour.

Such a ratio is special because, unlike the case of comparing lengths of two segments measured by the same unit of measurement (e.g., in cm), the ratio 25 resulted from comparing quantities measured by different units. In such case, the ratio is special because it compared quantities measured by different units. Put another way, the ratio between two quantities measured by different units is called a rate.

Another example of rate is to compare different currencies, e.g., US and Singapore Dollars. According to a currency converter (for July 7, 2021) for 100, 000 US Dollars one can get 134, 556 Singapore Dollars. The rate 134,556/100,000 can be reduced to 1.34556/1. The latter fraction represents the unit rate between the Singapore and the US currencies.

The concept of rate appears also in distance problems in terms of the average rate of speed with which different kinds of movement happen. History of mathematics preserved such problems being included in a number of sources. One source is the 15th century Italian book on arithmetic published by an unknown author the problems from which are reproduced in [Smith,1924]. Another source goes back to Liu Hui (225 – 295 A.D.), one of the greatest mathematicians of ancient China, and it deals with the goose and duck problem reproduced in [Dai and Cheung, 2015, p. 29].

As an example, consider

Problem 9.4. *The distance from A to B is 400 miles. It takes 8 hours to cover this distance by a car and 10 hours – by a truck. If the car and the truck started moving toward each other at the same time from A and B, respectively, in how many hours would they meet?*

Assuming that both vehicles move with a constant speed[25], the car driving from A to B will cover 400/8 = 50/1 miles in one hour. Likewise, the truck driving from B to A will cover 400/10 = 40/1 miles in one hour. The ratios 400/8 and 400/10 are not just ratios of miles to hours, they are special ratios referred to as the average rates of speed with which the car and the truck move, respectively. The rates of speed 50/1 and 40/1 are called unit rates as ratios with denominator one. To find the time it takes the vehicles to drive before they meet, one can construct an equation involving unit rates only: their sum, 50/1 + 40/1 = 90/1, represents the

[25] This assumption is true for all pre-calculus distance problems.

average rate of speed the vehicles move toward each other. Indeed, when the car covers 50 miles in one hour and, moving from the opposite direction, the truck covers 40 miles in one hour, they jointly cover 90 miles of the road during the same time; in other words, reducing the original distance between them by 90 miles. The unit rate 90/1 can be equated to another rate given by the ratio of 400 miles to the time x from the beginning of the movement till the meeting as, by the time the vehicles meet, the original distance between them would be reduced by 400 miles. Thus, one has to solve the equation $90/1 = 400/x$ whence $x = 40/9$ hours or 4 hours 26 minutes and 40 seconds. This conversion can be caried out by *Wolfram Alpha* as shown in Fig. 9.8.

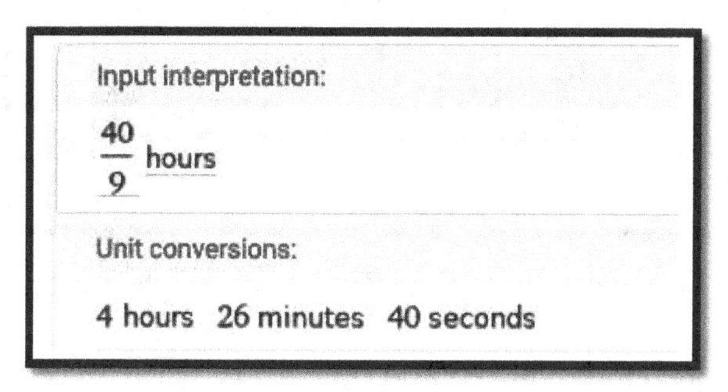

Fig. 9.8. Converting 40/9 hours in hours, minutes and seconds using *Wolfram Alpha.*

9.5.2 Solving the car and the truck problem using rectangular grids

Problem 9.4 can be solved using a rectangular grid. As always, using pictures as means of problem solving enhances one's conceptual understanding of the mathematics involved. In the course of solving such a distance problem on a grid, different ratio-related problems have to be solved. To begin, consider a 40-cell grid (Fig. 9.9), where each cell represents ten miles so that the entire grid represents the distance from A to B. Because the car and the truck move towards each other with the speed 50 and 40 miles per hour, respectively, they jointly cover 90 miles per hour or, on the grid, the car covers 5 cells and the truck covers 4 cells; jointly both vehicles cover nine cells in one hour. As shown on the grid of Fig. 9.9, because 36 of 40 cells include 9 cells exactly 4 times, after four hours the remaining distance between the car and the truck is 4 cells or 40 miles. Now one has to figure out how much time from that point the vehicles need in order to meet. If 9 cells are covered in 1 hour, then, under the assumption of uniform movement, one cell is covered in 1/9 hours and, consequently, 4 cells are covered in 4/9 hours. Alternatively, due to the uniform movement, the ratio of distance to time needed to cover it is a constant. Let the time (measured in hours) needed to cover 40 miles be equal to x. Therefore, $9/1 = 4/x$ whence $x = 4/9$ hours or 26 minutes and 40 seconds. In all, the time from the beginning of the vehicles' movement to their meeting is 4 hours, 26 minutes and 40 seconds.

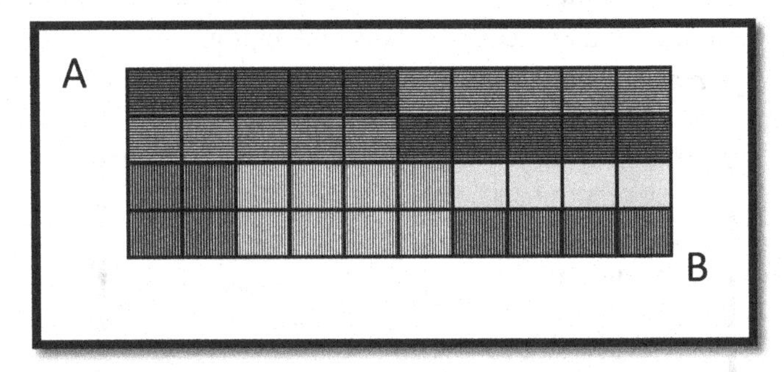

Fig. 9.9. Using a 40-cell grid to solve the car and the truck problem.

Consider now a 30-cell grid (Fig. 9.10). When 30 cells represent 400 miles, measuring 400 miles by 30 cells, as discussed in Section 9.1 of this chapter, one cell represents 40/3 miles. Because the car and the truck move towards each other with the speed 50 and 40 miles per hour, respectively, they jointly cover 90 miles per hour. Now, one has to find out how many cells represent 90 miles if one cell represents 40/3 miles. Solving the proportion $\frac{90}{x} = \frac{40/3}{1}$, where x is the number of cells representing 90 miles, yields $x = \frac{90 \cdot 3}{40} = \frac{27}{4}$ (cells). That is, both vehicles cover 27/4 cells in one hour. Measuring 30 cells by 27/4 cells leads to the ratio $\frac{30}{27/4} = \frac{40}{9}$. The fraction 40/9 represents the number of hours for both vehicles to drive towards each other before they meet. Thus, we have the same result as before – 4 hours 26 minutes and 40 seconds.

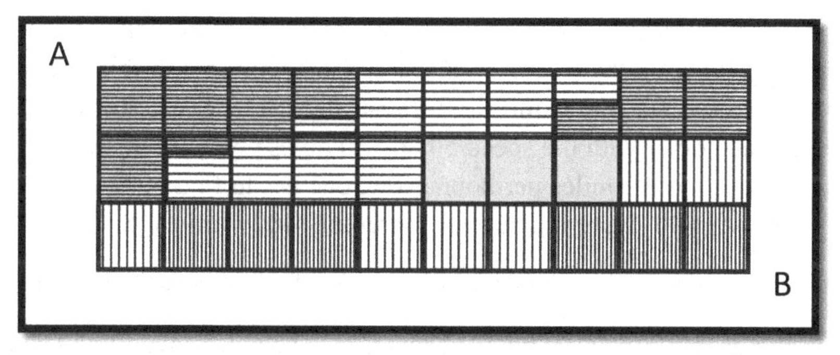

Fig. 9.10. Using a 30-cell grid to solve the car and the truck problem.

Consider now a 20-cell grid (Fig. 9.11). When 20 cells represent 400 miles, measuring 400 miles by 20 cells, as discussed in Section 9.1 of this chapter, one cell represents 20 miles. Because the car and the truck move towards each other with the speed 50 and 40 miles per hour, respectively, they jointly cover 90 miles per hour. Now, one has to find out how many cells represent 90 miles if one cell represents 20 miles. Solving the proportion $\frac{90}{x} = \frac{20}{1}$, where x is the number of cells representing 90 miles, yields $x = 90/20 = 9/2$. That is, both vehicles cover 9/2 cells in one hour. Measuring 20 cells by 9/2 cells leads to the ratio $\frac{20}{9/2} = \frac{40}{9}$. The fraction 40/9 represents the number of hours for both vehicles to drive towards each other before they meet. Thus, we have the same result as before – 4 hours 26 minutes and 40 seconds.

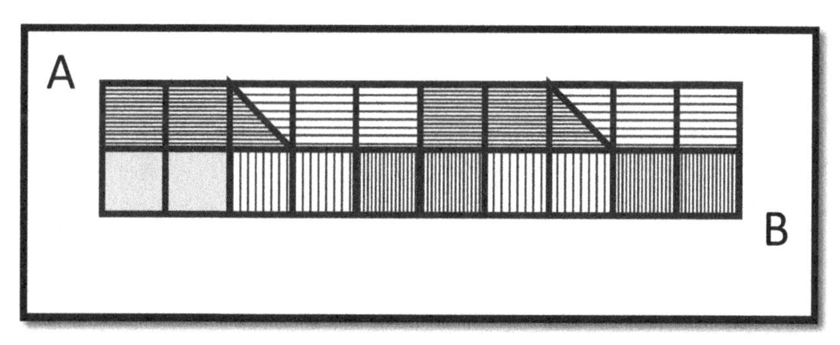

Fig. 9.11. Using a 20-cell grid to solve the car and the truck problem.

Finally, consider an n-cell grid. When n cells represent 400 miles, measuring 400 miles by n cells, as discussed in Section 9.1 of this chapter, one cell represents $400/n$ miles. Because the car and the truck move towards each other with the speed 50 and 40 miles per hour, respectively, they jointly cover 90 miles per hour. Now, one has to find out how many cells represent 90 miles if one cell represents $400/n$ miles. Solving the proportion $\frac{90}{x}=\frac{400/n}{1}$, where x is the number of cells representing 90 miles, yields $x=90n/400=9n/40$. That is, both vehicles cover $9n/40$ cells in one hour. Measuring n cells by $9n/40$ cells, leads to the ratio $\frac{n}{9n/40}=\frac{40}{9}$. The fraction 40/9 represents the number of hours for both vehicles to drive towards each other before they meet. Thus, regardless of the number of cells in a grid used to solve Problem 9.4, we arrive at the same result as before – 4 hours 26 minutes and 40 seconds.

CHAPTER 10: GEOMETRY

10.1 Basic concepts of geometry

10.1.1 Triangle

Any three non-collinear points (i.e., points that do not belong to the same straight line) uniquely define a triangle, the construction of which consists of connecting the points – vertices of the triangle, with line segments – sides of the triangle. It is interesting to note that any three non-collinear points also determine a unique plane to which the triangle belongs. Three types of triangles can be considered: an equilateral triangle – all three sides have equal lengths, an isosceles triangle – only two sides have equal lengths, and a scalene triangle – all three sides have different lengths. Whereas any three non-collinear points uniquely define a triangle, not any three segments may serve as sides of a triangle. Only when the combined lengths of *any* two segments is greater than the length of the third segment, a triangle can be constructed out of the three segments. This relationship among the three side lengths, known as the triangle inequality, is intuitively clear, yet its formal proof is complicated and beyond the scope of the pre-college mathematics curriculum. It is through intuition that one often cuts across the grass rather than walks over the pavement to get to a building faster. In doing so, one uses the triangle inequality – cutting across the grass takes a segment which is shorter than two (often mutually perpendicular) segments delineated by the pavement.

Another interesting property of a triangle is that the sum of its three internal angles measures by the straight angle, 180°. This fact can be demonstrated by using informal geometry (although its formal proof is not beyond the scope of the pre-college mathematics curriculum). To this end, one can take a bank paper, fold it as if it should be put into a business envelope, draw a triangle (using a ruler) on the so folded paper, numerate its angles and then cut the folded paper along the sides of the triangle to have three identical triangles. Each copy should then be marked with identical labeling of angles. After that, the three identical copies have to be arranged as shown in Fig. 10.1. Now, one can check to see that the three triangles form a trapezoid providing visual evidence that the sum of the three angles is a straight angle. In Section 10.5 this informal geometry activity will be extended to tessellation. The use of the *Geometer's*

Sketchpad in the construction of an equilateral triangle, a square, and other regular polygons is discussed in Chapter 13.

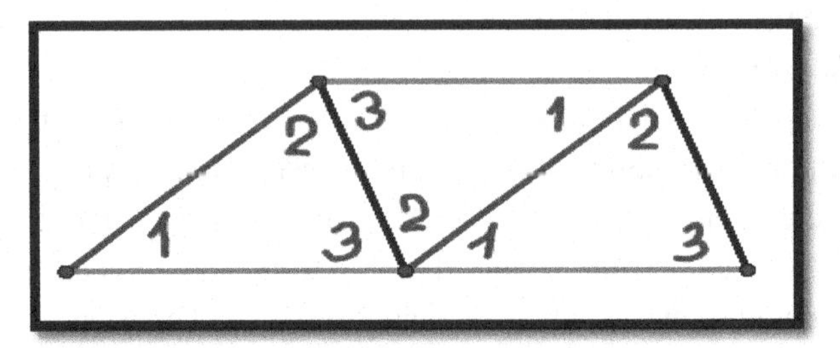

Fig. 10.1. Measuring three angles of a triangle.

10.1.2 Rectangle

Rectangle can be defined as a quadrilateral with four right angles. Basic geometric activities with a rectangle can be associated with different ways of dividing the figure in two identical parts. Fig. 10.2 shows six ways of doing that. The sketches demonstrate different ways of contextualizing from the abstract symbol 1/2 and show how geometric figures of the same area may have different perimeters. One may be asked to find out which shape has the largest perimeter without actually calculating perimeter but through comparing lengths of the corresponding segments. The statement about equal areas is due to a simple deduction; namely, if two figures are identical, they have equal areas (a numeric characteristic of a plane figure to be determined below) and, therefore, halves of these figures have equal areas as well.

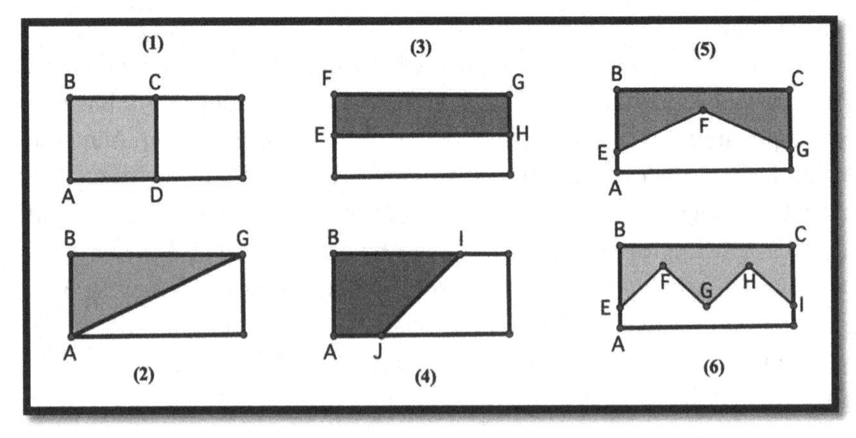

Fig. 10.2. Different ways to cut a rectangle in half.

Assuming that we deal with an integer-sided rectangle, in some cases (e.g., Fig. 10.2, cases 2, 4, 5, 6), perimeter of a shaded polygon cannot be found by measuring the side lengths by the linear unit used to measure side lengths of the rectangle (except for special values of the side lengths of a rectangle). For example, a diagonal of the rectangle, in general, cannot be measured by a linear unit (Fig. 10.2, case 2). In some cases, like for a three by four rectangle (Fig. 10.3), its diagonal includes the linear unit five times. An interesting activity is to construct on a geoboard (Section 10.4) a three by four rectangle to see that its diagonal can be measured by either of its sides (see Chapter 9, Section 9.1). Another such rectangle has side lengths measured by six and eight linear units (noting that $6 = 2 \cdot 3$, $8 = 2 \cdot 4$; thus, the side lengths of the three by four and six by eight rectangles are in the same ratio). The construction of the former rectangle, constructing its diagonal and measuring it by the linear unit is shown in Fig. 10.3. By drawing a circle centered at point D (one of the vertices of a three by four rectangle) with the unit radius AB, drawing the diagonal DF, marking with letter G the intersection of the diagonal and the circle, making a vector DG, and translating point G by this vector, one can see that diagonal DF includes the linear unit AB exactly 5 times.

Looking for other rectangles with a different ratio of the side lengths and experiencing difficulties in finding such rectangles can motivate students to learn about Pythagorean triples and the associated Pythagorean theorem. Entering the words "Pythagorean triples" into the input box of *Wolfram Alpha* allows one to see Pythagorean triples in the form of the Pythagorean theorem and to note that the largest element of such a triple can be measured by either of its other two elements (Fig. 10.3) and when the largest element is used as the length of the diagonal of a rectangle, the latter has other two elements of the triple as the side lengths of the corresponding rectangle. Furthermore, one can be asked to determine the side lengths of a rectangle and the location of other points, both on the border and inside the rectangle, so that perimeters of its halves shown in Fig. 10.2, cases 5 and 6, are whole numbers.

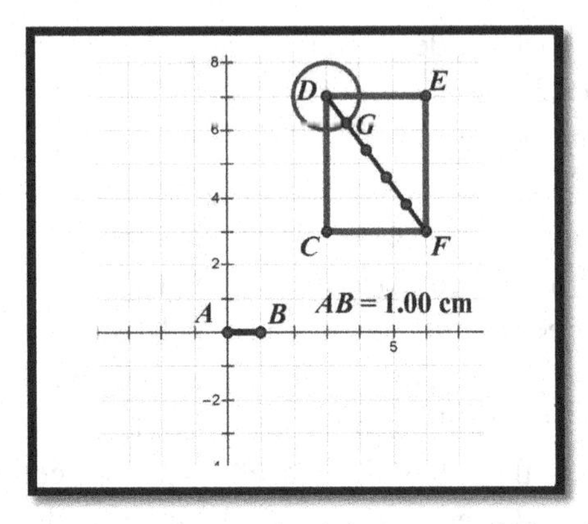

Fig. 10.3. Measuring a diagonal of the rectangle by the linear unit.

10.2 The van Hiele levels of geometric thinking

Pierre van Hiele and Dina van Hiele-Geldof, Dutch mathematics educators, in the mid of the 20th century proposed a theory of how the learning of geometry occurs by students moving from one level of geometric thinking to another (higher) level. The theory [van Hiele, 1986] suggests five such levels, labeling them by the numbers 0, 1, 2, 3, and 4. Level 0, called visualization, is the basic level when students can only recognize geometric shapes by their appearance without making any conceptual analysis in order to put several seemingly different shapes in a single group or decide whether the shapes belong to different groups. At Level 0, there is only one group, called shapes, as young learners of mathematics do not see any difference between various quadrilaterals and sometimes even may not see a difference between a circle and a triangle. Those are just shapes, enclosing space in the plane. For example, if the sides of a triangle are not rigid, it can be transformed into a circle (and vice versa), thus pointing out to the fact that both figures have two parts – the inside and the outside of an enclosure. This perception of shapes also points at the so-called topological primacy in spatial thinking [Piaget and Inhelder, 1963]. The term stems from a branch of mathematics, called topology, which studies continuous (i.e., without tearing) modifications of geometric figures such as stretching, twisting, and bending under the assumption of pliability of their boundaries.

The next level of thinking is Level 1, called analysis. At that level students are able to classify polygonal shapes (triangles, quadrilaterals, pentagons) by the number of sides, distinguish between convex and concave shapes, identify shapes with curvilinear borders, recognize (first visually and then through measuring) regular polygons (i.e., polygons with equal side lengths and same measures of angles), and so on. Note that a rhombus has equal side lengths, but its angles are not all measured the same; therefore, only square is a regular quadrilateral. In the primary grades, the main role of a teacher in teaching geometry is to help students move from Level 0 to Level 1. As mathematics educators in South Africa suggest, "in geometry, it is no longer sufficient to focus on what a shape looks like, they [students] need to focus on the properties of shapes" [Department of Basic Education, 2018, p. 58]. Activities may include sorting pattern blocks, using the blocks to build a figure, comparing and contrasting figures built out of blocks (Section 10.3).

Level 2 is called informal deduction. It is used by teachers both in a lower and in an upper elementary classroom. A simple example of informal deduction appropriate for young children is the use of paper and scissors in demonstrating that a diagonal of rectangle divides it in two identical triangles (see Chapter 7, Section 7.1, Remark 7.1), because paper folding – another strategy of informal deduction that works in the case of demonstrating line symmetry – does not work in that case. Examples of informal deduction appropriate for upper elementary grades include the derivation of Pick's formula (Section 10.4), the discovery that all triangles and quadrilaterals tessellate (Section 10.5) and identifying square as having the smallest perimeter or the largest area among all rectangles with a given area or a given perimeter, respectively (Section 10.7). All these informal deduction probes are consistent with "an intuitive and experimental approach ... [for students] to explore geometric shapes and properties through hands-on activities" [Ministry of Education Singapore, 2020, p. 20].

Level 3, called formal deduction, is related to high school mathematics education. An example of geometric thinking at that level is the use of the side-side-side property of the congruency of triangles (i.e., having identical size and shape so that they can be precisely overlaid) when proving that a diagonal cuts rectangle in two congruent triangles.

One can say that all geometric properties taught in high school require the use of formal deduction.

Finally, the highest level of geometric thinking, Level 4, is called rigor. It relates to the tertiary studies of geometry by mathematics majors. At that level, students learn different axiomatic geometries. For example, they learn that all properties of geometric shapes in Euclidean geometry studied in high school follow from the axiom (a statement that belongs to the foundation of a certain theory and is taken as true without any proof) that through a point outside a straight line, one can draw one and only one straight line which is parallel to the first one. For many years, without success, mathematicians tried to turn this axiom into a theorem (for the fewer axioms belong to the foundation of a theory, the more cogent the theory is) until the opposite was assumed, namely, that through a point outside a line, one can draw more than one line which is parallel to the first one. This revolutionary decision, associated mostly with the names of Lobachevsky[26] and Bolyai[27] who, independently of each other, proposed to negate the axiom, resulted in the creation of other geometries, different from Euclidean geometry.

10.3 Basic activities at Level 1 of the van Hiele model

In order to help young children to proceed from Level 0 to Level 1 in their geometric thinking, different activities with pattern blocks can be carried out by teachers. These activities can be integrated with the ideas of patterning and sorting the blocks. Consider Fig. 10.4. The first line of pattern blocks alternates two different blocks representing triangles and squares. To answer the question "What comes next?", a child has to distinguish between triangle and square on the very basic level at which the analysis includes only visual distinction without counting the number of sides. The recognition of the emerging pattern shown in the second line requires the integration of visualization and counting; but counting does not involve geometric analysis – it is the growing quantity of two visually distinct shapes that matters. The third line involves three different shapes in a more complicated repetition, something that prepares a child to

[26] Nikolai Ivanovich Lobachevsky (1792-1856) – a Russian mathematician, notable contributor to the advancement of university education and public enlightenment.

[27] János Bolyai (1802-1860) – a Hungarian mathematician.

recognize a shape not only from a visual standpoint, but also to see all shapes from a directional (or counting up and down) perspective and to wonder whether repetition can be developed in more than one way. Finally, the fourth line of Fig. 10.4 represents the sequence of shapes in which one must recognize their geometric distinction in terms of the number of sides. Different continuations of this sequence of shapes may be offered, one of which can be informed by the shapes in the third line. That is, now the child has to count not the shapes but their sides as the basic characteristics of the shapes. The above activities suggest that geometric thinking which is indicative of Level 1 of the van Hiele model can, nonetheless, be observed in a young learner when he or she is appropriately guided by a teacher, who should assume the possession of such skills by a child at the rudimentary stage of development.

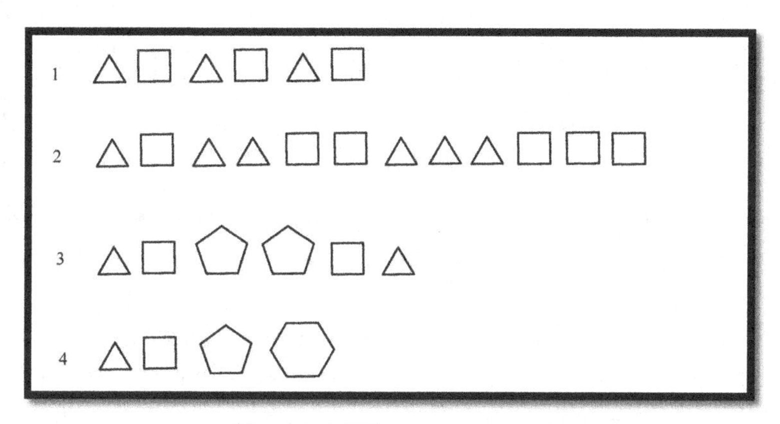

Fig. 10.4. What comes next?

More complicated activities with pattern blocks are presented in Fig. 10.5. Here, a teacher can ask a student to describe a method used in decomposing four blocks – equilateral triangle (ET), square (SQ), regular hexagon (RH) and isosceles trapezoid (IT) – in two groups. This task goes beyond visual recognition of differences among the four shapes and requires skills in geometric thinking. The first decomposition shows the distinction between quadrilaterals and not quadrilaterals. The second, third, fourth and fifth decompositions show the distinction between, respectively, ET and not ET, SQ and not SQ, RH and not RH, IT and not

IT. In particular, one can see that there are four ways to select an object from a set of four different objects (see Chapter 11, Section 11.5). The third decomposition can also be interpreted as including in the group of three objects those into which the RH can be partitioned (RH = 2 IT = 6 ET), yet partitioning of RH into the shapes available would not yield SQ. Finally, the sixth decomposition may be interpreted as regular polygons vs. a not regular polygon.

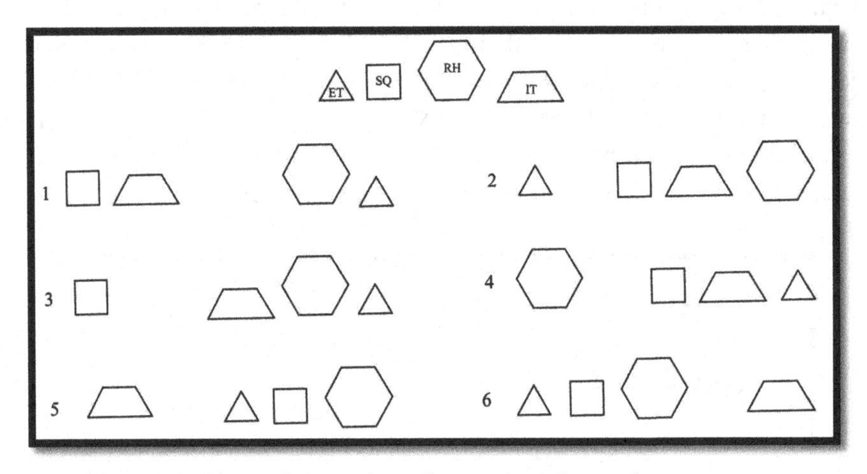

Fig. 10.5. The task is to describe sorting shapes in two groups.

Regarding teacher-student interaction through the activities discussed in this section, the following quote from Vygotsky [2001, p. 84, translated from Russian by the author], whereas related to the development of reasoning about number and arithmetic, is worth mentioning in the context of geometry: "*Although at an early stage of mathematical development, quantitative reasoning and arithmetical thinking of a child are pretty vague and immature in comparison with those of an adult with whom the child interacts, it is through this interaction that the final forms of reasoning and thinking about numbers, that have to be developed as a result of having an adult in his/her environment, are somehow present at that stage and, not only present but in fact define and guide the child's first steps toward the development of the final forms of understanding quantity and comprehending arithmetic*". Paraphrasing Vygotsky, one can say that although at an early stage of mathematical development, spatial reasoning and geometric thinking of a child are pretty vague and immature

(Level 0) in comparison with those of a teacher with whom the child interacts, it is through this interaction that the final forms of reasoning and thinking about shapes, that have to be developed as a result of having the teacher in the child's environment, are somehow present at that stage and, not only present but in fact define and guide the child's first steps toward the development of the final forms (Level 1) of spatial understanding and comprehending properties of geometric figures. In other words, the role of a teacher in helping a child to move from Level 0 to Level 1 is critical.

10.4 Geoboard activities and Pick's formula

A geoboard is a hands-on learning environment used for exploring basic geometric ideas associated with different polygonal shapes [Gattegno, 1971]. It was introduced into mathematics education by Caleb Gattegno, a British mathematics educator, known also for promoting the use of Cuisenaire rods. A geoboard environment allows for the construction of a variety of polygons using rubber bands held by pegs. A polygon on a geoboard can be associated with the number of pegs that the rubber band touches. Another characteristic of a polygon on a geoboard is the number of pegs in its interior. Thus, polygons on a geoboard may be compared in terms of the numbers of two types of pegs associated with them. For example, in one case, a rubber band touches six pegs and encloses three pegs; in another case, a rubber band touches eight pegs and encloses two pegs. As will be shown below, the two polygons have equal areas (a concept to be defined below). In a mean time, without knowing how to find area, one can explore whether on a geoboard both polygons could be triangles.

This association of polygons, both convex and concave, with pegs on a geoboard brings about counting as one of the major problem-solving strategies furnishing geoboard geometry with an informal flavor, something that is especially important at the elementary level. At the same time, appropriately designed counting activities on a geoboard can be conceptually rich and the exploration suggested at the end of the last paragraph is such an example. Furthermore, because on a geoboard a linear unit is a side of a unit square the vertices of which are located at four pegs closest to each other, one can find area of any shape by using a strategy shown in Fig. 10.6. This strategy consists of enclosing the shaded shape into a rectangle (square) and then subtracting from its area the (easy to find) areas of extraneous triangles as they always have half of area of a rectangle which, in turn, is comprised of unit squares. This is due to the fact, which can be confirmed through informal geometry (see Section 10.2) using paper and scissors, that a diagonal of a rectangle cuts the latter in two congruent triangles. In other words, addition, subtraction, and dividing integers by two are three major arithmetical operations in the context of finding areas on a geoboard. Furthermore, the areas on a geoboard are always multiples of one-half of area of the unit square; in other words, area on a geoboard is always an integer divided by two.

In the United States, expectations of Common Core State Standards [2010, p. 40] for students in Grade 6 include finding "areas of right triangles, other triangles, and special quadrilaterals by decomposing these shapes, rearranging or removing pieces, and relating the shapes to rectangles". Likewise, in South Africa, students learn "to calculate the area of complex shapes by breaking down the shape into familiar shapes like rectangles and triangles" [Department of Basic Education, 2018, p. 53]. In Japan, students are taught that "the areas of complex figures may be calculated more easily using the formulas for areas of squares and rectangles" [Takahashi et al., 2004, p. 221]. Note that in early grades, students are not using formulas for areas; instead, especially on a geoboard, they are measuring areas by unit squares.

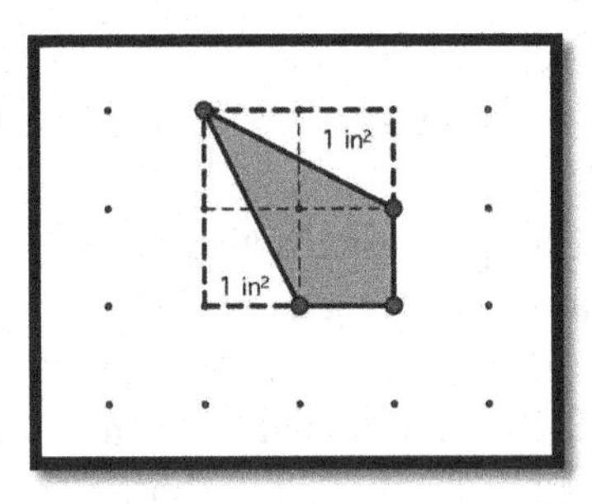

Fig. 10.6. Finding area of the shaded polygon using informal geometry.

Such informal approach to geometry was investigated in the mid 1950s in New York city public schools by Max Wertheimer, one of the founders of Gestalt psychology, and his collaborators. Their investigation concerned the use of productive thinking in problem solving [Wertheimer, 1959] involving children as young as five-year-old in finding areas of rectangles, parallelograms, isosceles triangles, and trapezoids. Reflecting on this research, Luchins and Luchins [1970, p. 44] acknowledged historical roots of informal geometry, an instructional approach used by the modern-day teachers worldwide, in the context of geoboards: "Wertheimer's method of finding the areas of geometric figures by transforming them into others, particularly rectangles and triangles, is an ancient one. It has been used, for example, in problems given in an Egyptian papyrus roll [dated circa 1650 B.C.], found in 1858 by [a Scottish scholar and collector of antiques] Henry Rhind". In particular, these methods have been used by elementary school teachers in Japan when helping students to understand "that the idea for finding the area of a triangle using the equivalent-area transformation of the rectangle into a parallelogram by bisecting the height can be used to determine the area of special triangles and trapezoids" [Takahashi et al., 2004, p. 263]. So, the notion of diversity of mathematics teaching methods includes not only their international aspect but also the sagacity of Egyptian civilization going back to several millennia in the history of human race.

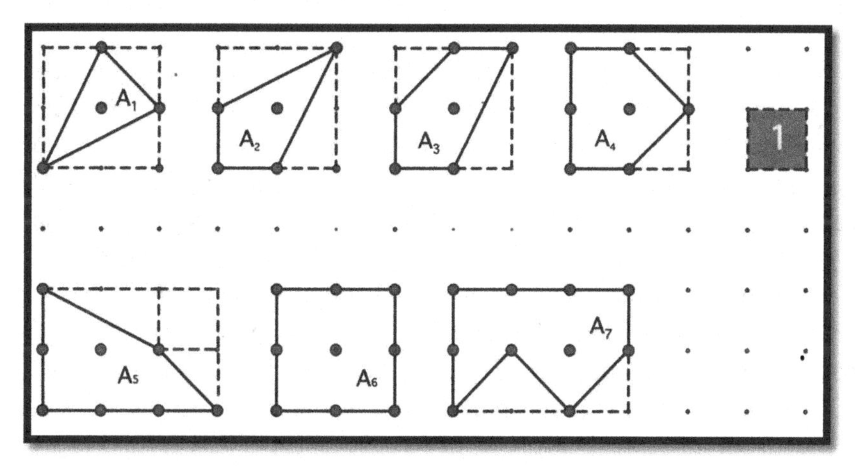

Fig. 10.7. Finding areas of the polygons with $I = 1$ and $B = 3, 4, \ldots, 9$.

Having in mind educational significance of the above-mentioned historical connection, consider seven polygons with solid borders shown in Fig. 10.7. One can see that the first four polygons have their areas increased by 1/2 as the number of their border pegs increases by one and there is no increase in the number of internal pegs. Indeed, using the strategy shown in Fig. 10.6 yields

$$A_1 = 4 - \left(1 + 1 + \frac{1}{2}\right) = \frac{3}{2}, A_2 = 4 - (1 + 1) = \frac{4}{2}, A_3 = 4 - \left(1 + \frac{1}{2}\right) = \frac{5}{2},$$
$$A_4 = 4 - \left(\frac{1}{2} + \frac{1}{2}\right) = \frac{6}{2}.$$

The same can be said about the three polygons in the bottom part of Fig. 10.7. Indeed,

$$A_5 = 6 - \left(1 + 1 + + \frac{1}{2}\right) = \frac{7}{2}, A_6 = 4 = \frac{8}{6}, A_7 = 6 - \left(\frac{1}{2} + \frac{1}{2} + \frac{1}{2}\right) = \frac{9}{2}.$$

One can see that the numerator in a fraction representing area of each of the seven polygons is equal to the number of pegs (B) on their borders, respectively, and all the polygons have a single internal peg (I), the presence of which does not affect the area. From this observation, one can conjecture that when $I = 1$, area of a polygon with B border pegs and one internal peg, $A(B, 1)$, can be computed through the formula $A(B, 1) = B/2$.

Fig. 10.8 shows a different situation. The first four polygons with four border pegs ($B = 4$) and one internal peg ($I = 1$) have $A(\mathrm{B}, 1) = 4/2 = 2$ as they are similar to those shown in Fig. 10.7. The next group of four polygons with four border pegs ($B = 4$) and two internal pegs ($I = 2$) have area $A(B, 2) = 3 = B/2 + 1$. The last group of four polygons with four border pegs ($B = 4$) and three internal pegs ($I = 3$) have area $A(B, 3) = 4 = B/2 + 2$. One can conjecture that the second term in the formulas for area is one smaller than the number of internal pegs. Thus, one can complete the conjecture by writing down the following formula (known as Pick's formula[28]) for area of a polygon on a geoboard

$$A(B, I) = \frac{B}{2} + I - 1. \qquad (10.1)$$

---

[28] George Alexander Pick (1859-1942), an Austrian mathematician who died at the age of 83 in the Nazi concentration camp.

A formal proof of Pick's formula (conjectured on the basis of a number of cases in favor of the formula; that is, through informal deduction) is complex and beyond the scope of the elementary school mathematics curriculum. Nonetheless, the above use of a geoboard is an example of what may be called activity-based learning which includes the use of "concrete manipulatives and experiences ... [proceeding from which] students are guided to uncover abstract mathematical concepts or results" [Ministry of Education Singapore, 2020, p. 17].

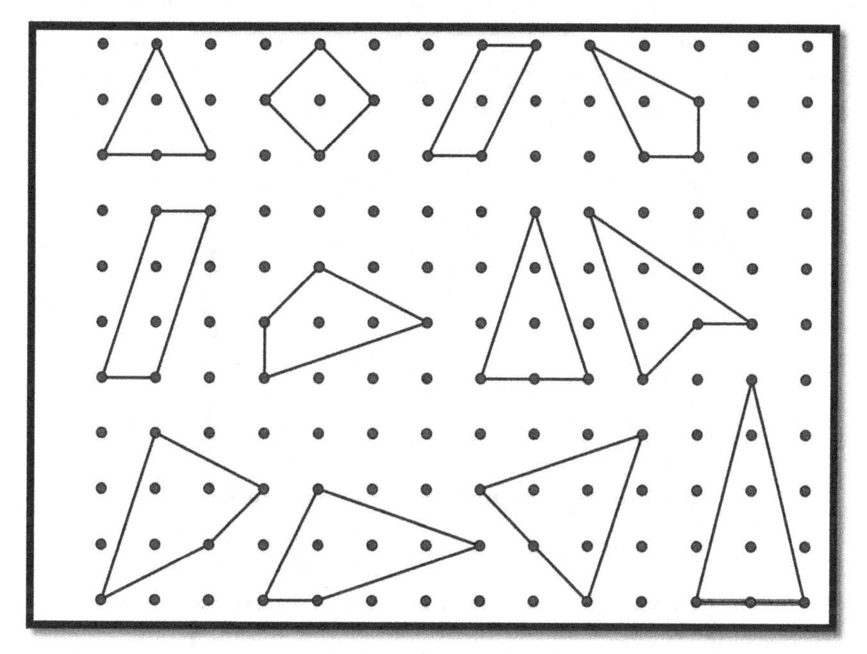

Fig. 10.8. Finding areas of the polygons with $B = 4$ and $I = 1, 2, 3$.

The context of Pick's formula can be used to integrate computational fluency with conceptual understanding. Consider the polygon pictured in Fig. 10.9. According to formula (10.1), its area is six square units. A conceptual part of dealing with Pick's formula is to explore ways of modifying the polygon to have area smaller/greater than (or even equal to) six. Fig. 10.10 shows how the polygon can be modified to have area seven square units. This modification does not change the number of internal pegs but, instead, it increases the number of border pegs by two. Fig. 10.11 shows another modification of the polygon of Fig. 10.9 by extending its size through replacing one border peg by another border peg and turning the replaced border peg into a new internal peg. This combination of the procedural and the conceptual supports the mathematics pedagogy of the 'single question – multiple answers' construct encouraging students' search for more than one correct solution to a problem [Abramovich, 2021], and confirms that "geometry learning provides opportunities to develop ability to reason mathematically" [Association of Mathematics Teacher Educators, 2017, p. 52]. Mathematical reasoning is a skill valued by educators in Chile by developing standards for teaching geometry "with an emphasis on discussion, reflection about possible hypotheses, alternative definitions and the use of counterexamples" [Felmer et al, 2014, p. 29].

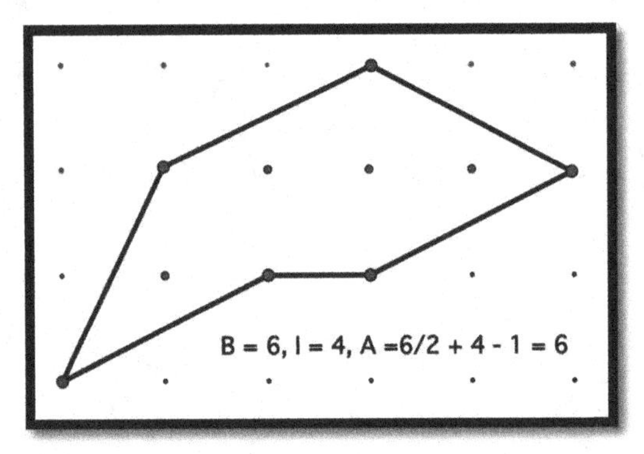

Fig. 10.9 Using Pick's formula to find area.

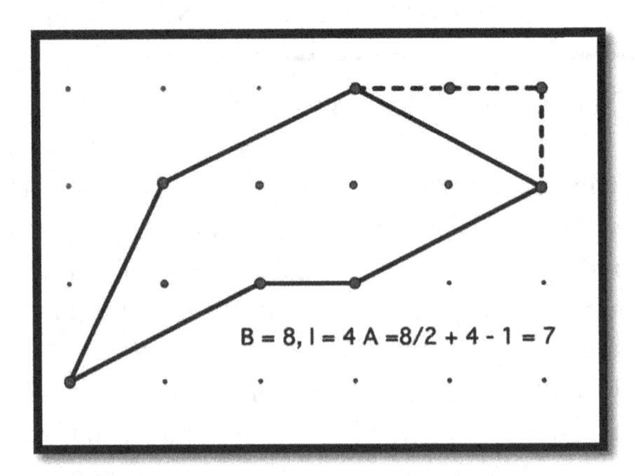

Fig. 10.10. Modifying polygon of Fig. 10.9 to have area 7 (square units).

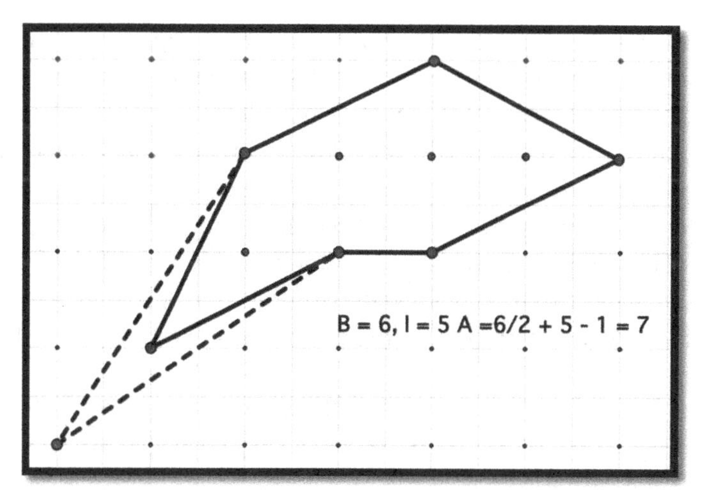

Fig. 10.11. Another alteration of Fig. 10.9 to have area 7 (square units).

10.5 Tessellation

As an introduction to the concept of tessellation, one can use a manipulative task borrowed from [New York State Education Department, 1998]: *Using pattern blocks such as green (equilateral) triangles, blue rhombuses, red (isosceles) trapezoids, and yellow (regular) hexagons, have students discover the quantity of green triangles needed to cover a blue rhombus, a red trapezoid, and a yellow hexagon.*

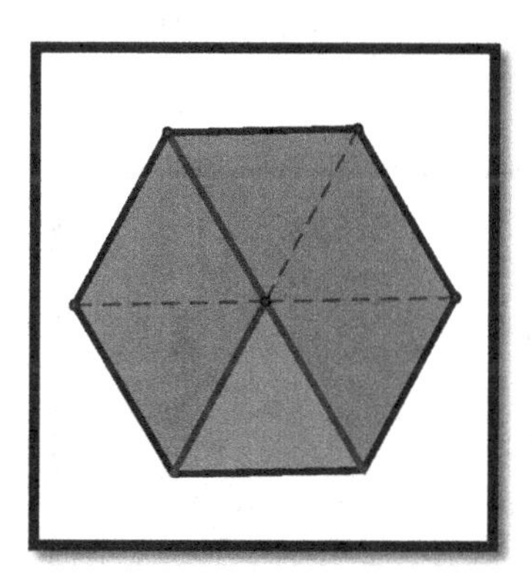

Fig. 10.12. Hexagon built from triangle, rhombus, and trapezoid.

Recommended as a geometric activity appropriate for Grades 1 and 2, this manipulative task has a hidden meaning. It enables rather sophisticated mathematical concepts (that elementary school teachers need to know) to be gradually developed. In that way, the task can motivate various mathematical activities. For example, Fig. 10.12 may be interpreted as covering space around a point with no gaps or overlaps using six identical equilateral triangles as shown in Fig. 10.13, because the rhombus includes two triangles and the (isosceles) trapezoid includes three triangles (see Section 10.3). In arithmetical terms, we have the following relations:

$$\frac{1}{2}+\frac{1}{3}+\frac{1}{6}=1, \frac{1}{2}=\frac{1}{6}+\frac{1}{6}+\frac{1}{6}, \frac{1}{3}=\frac{1}{6}+\frac{1}{6}.$$

The geometric construction shown in Fig. 10.13 can be extended to the whole plane by covering, step by step, space around any point using the triangles. In much the same way, one can cover the whole plane with trapezoids (not necessarily isosceles, as it will be explained later) and rhombuses. Such geometric activities are commonly referred to as tessellation: *covering the plane with either identical or different shapes in a repeated pattern with no gaps or overlaps*. Tessellations are studied in the 5th grade classroom in South Korea with an intent to teach geometry "in a deeper manner ... [motivating students] to make inference in geometric problem solving and reasoning at the elementary level" [Chang, 2013, p. 155].

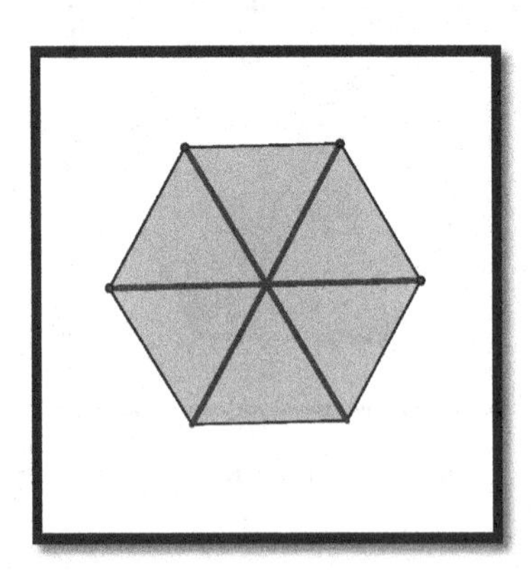

Fig. 10.13. Hexagon built out of six equilateral triangles.

For example, Fig. 10.12 shows a fragment of tessellation with three different polygons mentioned in the above manipulative task – an equilateral triangle, a rhombus (parallelogram with equal side lengths) and an isosceles trapezoid. This is a tessellation with three polygons two of which are not regular ones sharing at least one equal side length. Fig. 10.13 shows a fragment of tessellation with six identical equilateral triangles. Finally, a fragment of tessellation in which square, regular hexagon and regular dodecagon (a polygon with 12 sides) cover space around a point with no gaps or overlaps is shown in Fig. 10.14. One may note that $\frac{1}{4}+\frac{1}{6}+\frac{1}{12}=\frac{1}{2}$. This relationship among the reciprocals of the number of sides that square, hexagon and dodecagon have and the fraction 1/2 is not a coincidence. In general, the relation $\frac{1}{n}+\frac{1}{m}+\frac{1}{k}=\frac{1}{2}$ defines all integer values of n, m, and k to allow for tessellation with regular polygons having n, m, and k sides. There are ten integer triples (n, m, k) satisfying the last relation and the triple (4, 6, 12) is one of them. For more information on this topic see [Abramovich, 2010].

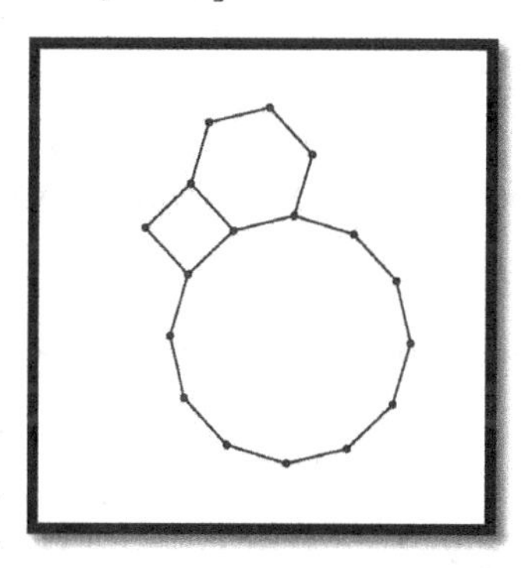

Fig. 10.14. A regular tessellation with square, hexagon, and dodecagon.

10.5.1 Tessellation with scalene triangles

As was shown in Fig. 10.13, by using six identical equilateral triangles one can cover space around a point without gaps or overlaps. The condition that triangles are identical is a critical one as equilateral triangles of different size (that is, similar equilateral triangles) would not provide the phenomenon of tessellation. An interesting activity is to explore if the condition of triangles being equilateral is needed for tessellation. To this end, a class may be asked to develop multiple sets of six identical scalene triangles by giving each student a bank paper, ruler, pencil, and scissors. By folding paper into three equal parts and then folding the so folded paper in half, one has six layers of paper. The next step is to follow the directions of Section 10.1.1 to get six identical scalene triangles with angles labeled as on the original triangle. Now, the task is to position the six triangles around a point in such a way that there are no gaps or overlaps (like in Fig. 10.1 in the case of three triangles). This arrangement of six scalene triangles is shown in Fig. 10.15. In this case, the methodology of informal deduction (Level 2 of geometric thinking in the van Hiele model introduced in Section 10.2) can be applied to conclude that all triangles tessellate (where the word *all* is limited to the number of students in the class). As we already discovered above (Fig. 10.1) the sum of three angles in any triangle is 180°, another conclusion which is due to informal deduction. This conclusion is another example of invariance in mathematics (see Chapter 1, Section 1.1, and Chapter 8, Section 8.1) – any variation of a triangle (along with its angles, something that can be demonstrated through an action using a dynamic geometry software) does not change the 180° property. The use of the *Geometer's Sketchpad* in demonstrating tessellation with scalene triangles is discussed in Chapter 13, Section 13.5.1.

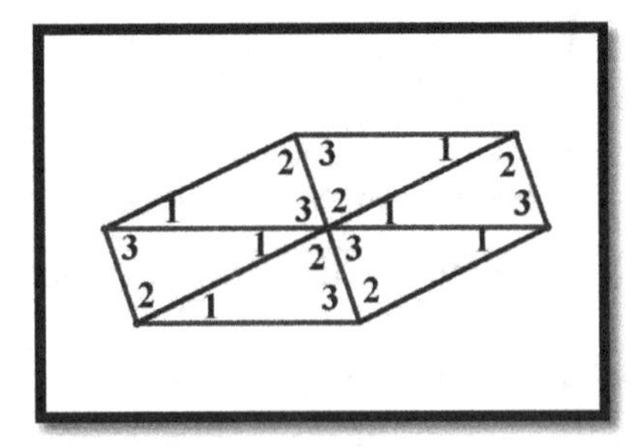

Fig. 10.15. Tessellations with scalene triangles.

10.5.2 Tessellation with quadrilaterals

A similar tessellation activity can be carried out with quadrilaterals. First, one can use four identical squares (or rectangles) to show an obvious outcome – four identical squares (or rectangles) can be used to cover space around a point with no gaps or overlaps. But as shown in Fig. 10.16, a set of four identical quadrilaterals covers space around a point as well. The condition of quadrilaterals being identical is critical for the continuation of tessellation by proceeding from other points. Once again, each student in a class can be given a bank paper which has to be folded in half twice and, like in the case of triangles, one has to use a ruler to draw a quadrilateral on the top of the folded paper, label its angles with the numbers 1, 2, 3, 4, then cut out four copies of identical quadrilaterals, and label angles on each copy exactly as on the original quadrilateral. Once each student in the class can make the arrangement of quadrilaterals shown in Fig. 10.16, the following conclusion, using the methodology of informal deduction (see Section 10.2), can be made: all quadrilaterals tessellate (where, once again, the word *all* is limited to the number of students in the class). Likewise, one can conclude that the sum of four angles in any quadrilateral measures 360°. This conclusion, just as in the case of triangles, is another example of invariance in mathematics (see Chapter 1, Section 1.1) – any variation of a quadrilateral (along with its angles, something that can be demonstrated through an action using a dynamic geometry software) does not change the 360° property. The use of the *Geometer's Sketchpad* in demonstrating tessellation with quadrilaterals is discussed in Chapter 13, Section 13.5.2.

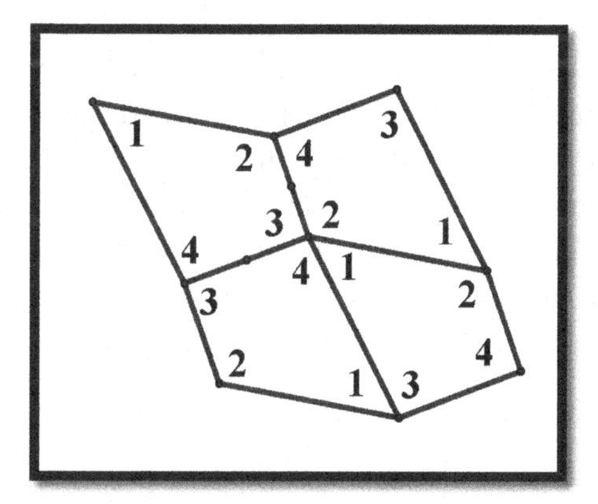

Fig. 10.16. Tessellation with quadrilaterals.

10.6 Creative thinking leading to special rectangles

When teachers are faced with a challenge presented by a student, they need to "not reject the challenge but to investigate the proposed idea, applying their own critical thinking and using all available resources" [Association of Mathematics Teacher Educators, 2017, p. 9]. Such experience can be acquired through the discussion and analysis of classroom episodes from the instructor's appropriately selected miscellany. Furthermore, according to the Conference Board of the Mathematical Sciences [2012, p. 23], elementary teacher candidates[29] "should study in depth the vast majority of K-5 mathematics, its connections to prekindergarten ... and grades 6-8 mathematics" in order to recognize the subject matter they are to teach being rich of "interesting ideas, which can be studied repeatedly, with ... attention to detail and nuance" [*ibid*, p. 31].

An example of intellectual courage required from a teacher to attend to the nuances of elementary mathematics can be provided through the reflection on the following classroom episode. An elementary teacher candidate gave to a second-grade student *eight* square tiles and asked the student to construct all possible rectangles out of the tiles. It was expected that the student would construct two rectangles as shown on the left-hand side of Fig. 10.17. The teacher candidate did not expect that the student would construct another rectangle having a hole as shown on the right-hand side of Fig. 10.17. The teacher candidate did not reject what was unexpected and, instead, praised the student for their creativity. In the Standards for Preparing Teachers of Mathematics [Association of Mathematics Teacher Educators, 2017, p. 22] one can find a note that sometimes, mathematics teachers when facing students whose thinking is different from what is expected, "may inadvertently seek to remedy those differences rather than seeing them as strength and resources upon which to build". A courage of not rejecting different from the mainstream mathematical thinking rendered by a student requires active participation in a course where numerous ideas are proposed and investigated making experience of teacher candidates educative due to its contribution to their

[29] The Conference Board of the Mathematical Sciences by the word "elementary" means K-5 level.

intellectual growth [Dewey, 1938]. Accepting ideas different from what is expected allows "students to really feel the importance of mathematical thinking and a rich sense of geometrical figures for creativity" [Takahashi et al., 2004, p. 261]. In much the same way, mathematics educators in South Africa believe that students "need to develop their ability to *think out of the box* (i.e., to find strategies that have not been shown to them before)" [Department of Basic Education, 2018, p. 18, italics in the original].

The third rectangle constructed by the second grader is not just a rectangle with a hole. It is special not only because it is a square, but because its area, 8, numerically, is half of its perimeter, 16 (if we consider the border of the hole as part of the perimeter, so that 16 = 12 + 4). One can check to see that the same relationship between area and perimeter continues for rectangles with a hole constructed from any even number of square tiles[30]. Furthermore, for any number of square tiles being a multiple of four greater than 8, more than one rectangle with a hole can be constructed. More specifically, out of $4n$ square tiles, $(n - 1)$ rectangles with a hole may be constructed, each having the perimeter equal to $8n$ linear units. When $n = 3$, out of 12 square tiles (Fig. 10.18) two rectangles with a hole can be constructed, one of which is a square; both having perimeter 24 linear units. The fact that out of $4n$ square tiles one can construct $(n - 1)$ rectangles with a hole having perimeter $8n$ is confirmed by the spreadsheet shown in Fig. 10.19 in the case $n = 5$. Note that this computational confirmation may not be accepted as a formal proof[31]. The spreadsheet (its programming details are included in Chapter 13, Section 13.11) can be used when posing problems about constructing rectangles,

---

[30] It appears that the teacher candidate chose to give the second grader eight tiles by accident. He could have given nine tiles to the child – in that case two traditional rectangles can be constructed only. Thus, it is due to serendipity that the child's hidden creativity was revealed through the construction of a rectangle with a hole.

[31] A proof requires one to note that when $4n$ tiles are arranged to have m and $2n - m + 2$ tiles, $3 \leq m \leq n + 1$, being a pair of adjacent sides of $n - 1$ $(= n + 1 - 3 + 1)$ rectangles, the perimeter of each such rectangle is equal to $2[(m + 2n - m + 2) + (m - 2 + 2n - m)] = 8n$. As an aside, the number of rectangles without a hole that can be constructed out of $4n$ tiles depends on the prime factorization of the number n.

including those with a hole and special relationship between area and perimeter. This allows one, as Japanese mathematics educators suggest, "to enrich the sense of geometric figures through activities that include observation and construction" [Takahashi et al., 2004, p. 75].

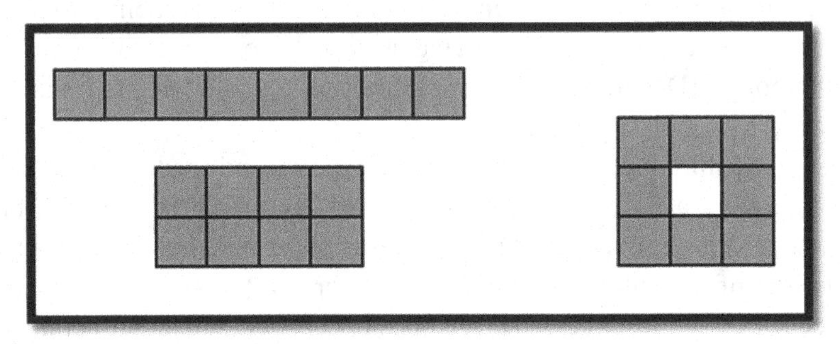

Fig. 10.17. A rectangle with a hole as a challenge for a teacher to accept.

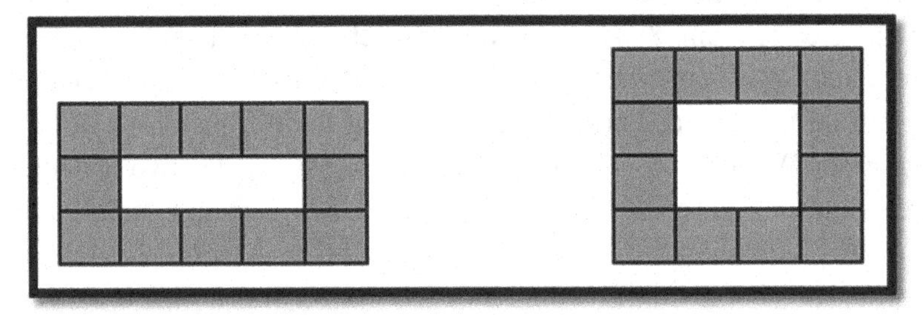

Fig. 10.18. Twelve square tiles afford two rectangles with a hole.

	A	B	C	D	E
1					
2	*n*	4*n*	Solutions	Area	Perimeter
3	5	20	4		
4		3	9	20	40
5		4	8	20	40
6		5	7	20	40
7		6	6	20	40

Fig. 10.19. There are four rectangles with a hole having area 20 and perimeter 40.

The second grader should be commended for clearly demonstrating creative thinking within a pretty mundane task. History of mathematics knows another type of rectangles with a special relation between area and perimeter – when area of a rectangle is numerically equal to its perimeter. There are only two such rectangles as it follows from the following classic quote attributed to Plutarch[32], "The Pythagoreans also have a horror for the number 17. For 17 lies halfway between 16 … and 18 … these two being the only two numbers representing areas for which the perimeter (of the rectangle) equals the area" [Van der Waerden, 1961, p. 96]. Perhaps the two rectangles, mentioned by Plutarch and known since the 6th century B.C., were also singled out due to someone's creative thinking about integer-sided rectangles. Note that mathematics educators in Australia suggest exploring shapes (not necessarily rectangles) "that have the same number of perimeter units as area units" [National Curriculum Board, 2008, p. 8]. Using a geoboard, one can check that right triangles with the side lengths (6, 8, 10) and (5, 12, 13) are such shapes.

10.7 On the relationship between perimeter and area of a rectangle

The Conference Board of the Mathematical Sciences [2012, p. 29] recommendations for mathematical preparation of elementary teacher candidates include the need to understand "the distinction and relationship between perimeter and area, such as by fixing a perimeter and finding the range of areas possible or by fixing an area and finding the range of perimeters possible". Understanding such distinction through problem solving by teachers is important as students already in the elementary classroom are expected to deal with "non-routine tasks that require deeper insights, logical reasoning and creative thinking" [Ministry of Education, Singapore, 2020, p. 10]. In Canada, teachers are expected to discuss with students the relationship between perimeter and area of a rectangle through the lens of "minimizing the amount of fencing needed; maximizing the area for a goat to graze" [Ontario Ministry of Education, 2020, p. 320]. Towards this end, the following activities, associated with Level 2

[32] Plutarch (46 A.D. – 119 A.D., Greece) – a historian of science.

(informal deduction) in the van Hiele model of geometric thinking can be carried out in the context of rectangles.

Activity 10.1. *Among all integer-sided rectangles of perimeter 24 cm, find the rectangle with the largest area and the rectangle with the smallest area.*

This activity is designed to approach conceptually area of rectangle as its fundamental characteristic. The first question to be addressed (although already addressed in the context of a geoboard, Section 10.4) is: How can one measure area of a rectangle? A not uncommon answer is when a teacher candidate provides the formula: "area is length times width". However, defining a concept through a formula can hardly be accepted as a conceptual approach to area of a geometric figure. To define area, one has to define a unit of measurement for area. This unit is called the unit square – a square of area one square unit. If we define a rectangle as a quadrilateral with the pairs of adjacent sides forming right angles, then an integer-sided rectangle with the side lengths n and m can be represented as an $n \times m$ array of unit squares, say, n unit squares going horizontally from left to right and m unit squares going vertically from top to bottom. But here, while one sees the product $n \times m$ as area, the product stems from counting unit squares and not from multiplying the side lengths. Likewise, the volume of a right rectangular prism with integer dimensions is understood as the number of unit cubes that such a prism includes. This is similar to measuring perimeter by using a ruler which shows a linear unit of measurement (e.g., 1 inch or 1 cm). So, both perimeter and area are computed by counting the corresponding units of measurement. This position is consistent with the Conference Board of the Mathematical Sciences [2012, p. 29, italics added] ideas about mathematics teachers' "understanding what area and volume are and giving *rationales* for area and volume formulas".

Having perimeter 24 cm, one has half of the perimeter representing the sum of the side lengths of two adjacent sides. Dividing 24 by 2 yields 12; that is, the sum of the lengths of two adjacent sides in the family of integer-sided rectangles of perimeter 24 cm is 12 cm. In order to find all rectangles that belong to the given family, one has to decompose the number 12 into a sum of two positive integers, a task already known to students from Grade 1 mathematics. In all, there are six decompositions:

$$12 = 11 + 1,\ 12 = 10 + 2,\ 12 = 9 + 3,\ 12 = 8 + 4,\ 12 = 7 + 5,\ 12 = 6 + 6.$$

That is, the pairs of adjacent sides are: (11, 1), (10, 2), (9, 3), (8, 4), (7, 5), (6, 6). The corresponding rectangles are shown in Fig. 10.20. By counting unit squares included in each rectangle, the values of areas (expressed in cm^2) are displayed to the left of each rectangle: 11 cm^2, 20 cm^2, 27 cm^2, 32 cm^2, 35 cm^2, 36 cm^2. One can see that rectangle with equal lengths of adjacent sides (i.e., square) has the largest area. One can try several other even number values for perimeters (in order to have the sum of lengths of two adjacent sides an integer) to confirm (empirically) that square has the largest area. Using informal deduction (Level 2 of the van Hiele model of geometric thinking), that is, making a general statement based on a number of examples in favor of generalization, one can conclude that, given a perimeter of the family of integer-sided rectangles, square always has the largest area and the "skinny" rectangle (that is, the rectangle with the unit side length) always has the smallest area.

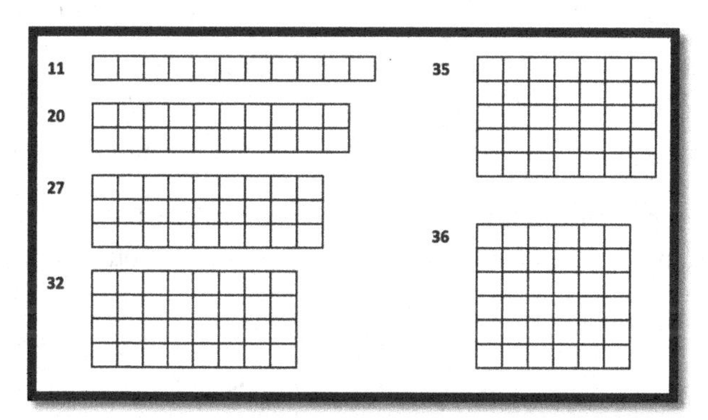

Fig. 10.20. Six integer-sides rectangles with perimeter 24 cm.

But what if we allow for a side length to be a non-integer; in particular, to be a number smaller than the number 1? In that case, one has to define area differently than through counting unit squares (if one wants to find area exactly, not to the accuracy of a whole number). It is at that point that counting unit squares in a rectangular array through multiplication *may be interpreted* as multiplying length by width, because the number of unit squares going from left to right is the length and the number of unit squares going from top to bottom is the width (or vice versa). In the case of the pairs introduced above for perimeter 24 cm (or semi-perimeter 12 cm) we have:

(11, 1) $\rightarrow A = 11 \times 1 = 11\ cm^2$; (10, 2) $\rightarrow A = 10 \times 2 = 20\ cm^2$; (9, 3) $\rightarrow A = 9 \times 3 = 27\ cm^2$; (8, 4) $\rightarrow A = 8 \times 4 = 32\ cm^2$; (7, 5) $\rightarrow A = 7 \times 5 = 35\ cm^2$; (6, 6) $\rightarrow A = 6 \times 6 = 36\ cm^2$;

In general, for a rectangle with the side lengths a and b, the area $A = a \times b$. Whereas the last formula is not rigorously proved, one can at least understand "where a mathematical rule comes from" [Common Core State Standards, 2010, p. 4]. Using this rule (i.e., formula), one can find area of the rectangle with perimeter 24 cm and the smaller side length equal to $\frac{1}{2}$ cm. We have $12 = \frac{1}{2} + 11\frac{1}{2}$ and, therefore, $A = \frac{1}{2} \times 11\frac{1}{2} = \frac{1}{2} \times \frac{23}{2} = \frac{23}{4} = 5\frac{3}{4}$. One can see that $5\frac{3}{4} < 11$; that is, by decreasing the smaller side length of a rectangle (and, consequently, increasing the larger side to keep perimeter the same), one decreases its area. This dynamic was already observed in the case of integer-sided rectangles of perimeter 24 cm. Likewise, $12 = \frac{1}{4} + 11\frac{3}{4}$ and, therefore, $A = \frac{1}{4} \times 11\frac{3}{4} = \frac{1}{4} \times \frac{47}{4} = \frac{47}{16} = 2\frac{15}{16} < 5\frac{3}{4} < 11$. By making the smaller side length as small as one wishes, the area of rectangle can also be made as small as one wishes; in other words, no rectangle with the smallest area can be found. For example, the decomposition $12 = \frac{1}{1000} + 11\frac{999}{1000}$ yields $A = \frac{1}{1000} \times 11\frac{999}{1000} = 0.01$ and such decrease in the value of area continues as long as one wishes. At the same time, among all rectangles with the same perimeter, square always has the largest area. This fact, mentioned by South African mathematics educators as an illustration that it is incorrect to "think that a square is not a rectangle" [Department of Basic Education, 2018, p. 73], can be proved at Level 3 in the van Hiele model of geometric thinking by using mathematical tools studied in high school.

Activity 10.2. *Among all integer-sided rectangles of area 36 cm^2, find the rectangle with the smallest perimeter and the rectangle with the largest perimeter.*

Now, by representing an integer-sided rectangle as a rectangular array of unit squares, one can find different rectangles with the given area through representing the number 36 as a product of two integers. We have: $36 = 36 \times 1$, $36 = 18 \times 2$, $36 = 12 \times 3$, $36 = 9 \times 4$, $36 = 6 \times 6$. That is, the pairs (36, 1), (18, 2), (12, 3), (9, 4), and (6, 6) represent adjacent side lengths of the family of integer-sided rectangles of area 36 cm^2. These rectangles with perimeters 74 cm, 40 cm, 30 cm, 26 cm, and 24 cm, respectively, are shown in Fig. 10.21. For example, $74 = 36 + 1 + 36 + 1$. Likewise, other perimeters can be calculated. Note that any integer, 36 included, can be factored in two factors one of which is not an integer in infinite number of ways, like $36 = \frac{1}{10} \times 360$, $36 = \frac{1}{100} \times 3600$, and so on. One can see that the rectangle with the side lengths represented by the pair (6, 6) has the smallest perimeter and the rectangle with the side lengths represented by the pair (36, 1) has the largest perimeter. That is, the rectangle with congruent adjacent sides (i.e., square) has the smallest perimeter, 24 cm; and the rectangle with the smallest possible integer side (a "skinny" rectangle) has the largest perimeter, 74 cm. However, already the factorization $36 = \frac{1}{10} \times 360$ corresponds to the rectangle with the side lengths $\frac{1}{10}$ cm and 360 cm, the perimeter of which is equal to $\frac{1}{10} + 360 + \frac{1}{10} + 360 = 720\frac{1}{5} > 74$. Likewise, the factorization $36 = \frac{1}{100} \times 3600$ yields a rectangle with the side lengths $\frac{1}{100}$ and 3600, the perimeter of which is equal to $\frac{1}{100} + 3600 + \frac{1}{100} + 3600 = 7200\frac{1}{50}$.

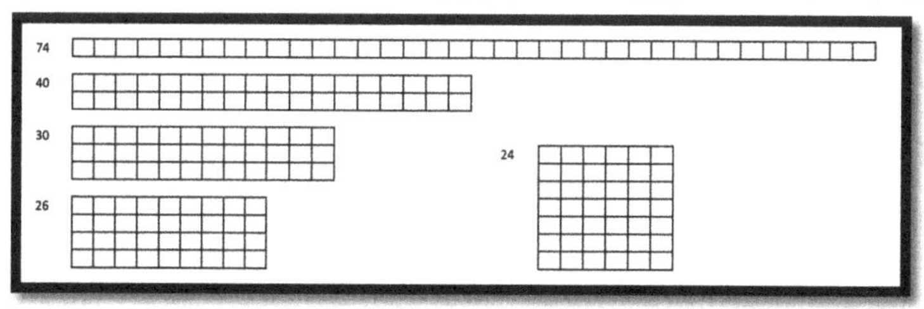

Fig. 10.21. Five integer-sides rectangles of area 36 cm^2.

That is, by allowing for a smaller side length of a rectangle of area 36 cm^2 (or any other area, for that matter) to be as small as one wants, perimeter of rectangle can be made as large as one wants. In other words, given area, no rectangle with the largest perimeter can be found. At the same time, by trying other (perfect square) values of area (to avoid dealing with irrational numbers) and using informal deduction (Level 2 of the van Hiele model of geometric thinking), one can conclude that given area, square always has the smallest perimeter. This geometric fact can be nicely applied to real life. Namely, if one looks for a rectangular piece of land with a given area and wants to minimize expenses for building a fence around the land, one has to buy a square piece of land. Likewise, as it was shown through Activity 10.1, when having a certain amount of fence, the largest rectangular piece of land to be fenced with this amount is the square. Put another way, one can recognize how the big idea of invariance in mathematics (see Sections 10.5.1, 10.5.2, and Chapter 1, Section 1.1) gives prominence to squares in the context of real-life applications.

CHAPTER 11: ELEMENTS OF COMBINATORICS: COUNTING THROUGH A SYSTEM

11.1 Tree diagrams

A tree diagram is a mathematical tool typically introduced to students in the third grade. It is used to organize counting according to a rule which can be explained through resolving the following real-life situation.

A cafeteria offers three types of drinks – water (w), coffee (c), tea (t) – and four types of fruit – apples (a), bananas (b), oranges (o), and pears (p). How many ways can one buy a pair of drink and fruit?

To answer this question by counting, one can start with creating possible (drink, fruit) pairs as follows: (w, a), (c, b), (t, p), (t, o), (c, a), and so on. Although "representing names concisely by using symbols is effective for systematic sorting and organizing" [Takahashi et al., 2004, p. 319], by listing such pairs in that way, one does not follow any rule (system) which would serve as an acceptable justification (proof) that all pairs have been found. Systematic reasoning through 'sorting and organizing', a forerunner of mathematical proof, has to be taught as this is not something that one intuitively possesses. For example, if a child is asked to write down all counting numbers smaller than 10, the child might think and write as follows: well, I remember that 5 is smaller than 10, and 7 is also smaller than 10, so are the numbers 2 and 3. From here, the list 5, 7, 2, 3, ... might emerge; yet it does not follow any rule other than uses immature memory. But a child could be taught to use basic counting skills (Chapter 1) and start counting from one obeying the order of numbers (without repeating them), so that the following (systematic) list would emerge: 1, 2, 3, 4, 5, 6, 7, 8, 9. In creating this list of numbers smaller than 10, one uses a system (rule) by starting with the smallest (counting) number and finishing with the largest (counting) number smaller than 10 through the process of counting by ones.

With this in mind, returning to the cafeteria story, one can first list all pairs with water by adding to water all fruits that appear in order on the fruit list. In that way, the pairs (w, a), (w, b), (w, o), (w, p) can be developed. Next, one can replace water with coffee in the above four pairs to get new pairs: (c, a), (c, b), (c, o), (c, p). Finally, the first letter in each pair can be replaced by the letter t to have (t, a), (t, b), (t, o), (t, p). One can see that pairs with fruit were repeated three times; that is, repeated as many times as the number of drinks available. One can see the emergence of repeated addition here; an arithmetic procedure known as multiplication. Recall, that in Chapter 2, Section 2.3, multiplication was introduced through counting cookies that were put in five boxes, each box having six cookies. (That is, six cookies were repeated five times). In the cafeteria situation, we have *four* fruits repeated *three* times. That is, the total number of the (drink, fruit) pairs can be expressed through the product 3×4. Put another way, the drink can be selected in three ways and, following this selection, the fruit can be selected in four ways, so that the ordered pair (drink, fruit) can be selected in 3×4 ways. In general, this leads to

The Rule of Product. *If an object A can be selected in m ways, and, following this selection, an object B can be selected in n ways, then the ordered pair (A, B) can be selected in $m \times n$ ways.*

Now, all 12 (drink, fruit) pairs can be represented through the diagram of Fig. 11.1. The diagram, "used to sort data into all the possible combinations of characteristics for two or more attributes" [Ontario Ministry of Education, 2020, p. 204], resembles a tree with branches connecting drinks with fruits and it is called a tree diagram. Just as the factors in the product 3×4 can be swapped to have the product 4×3, one can first select a fruit (in four ways) and then select a drink (in three ways) to have an equivalent tree diagram representation shown in Fig. 11.2.

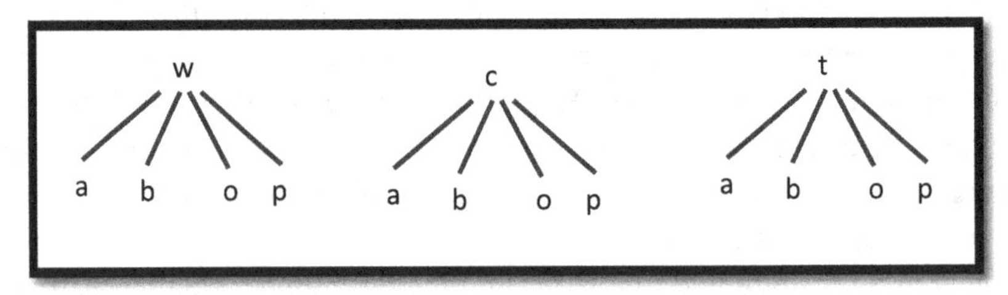

Fig. 11.1. A tree diagram representation of the drink/fruit situation.

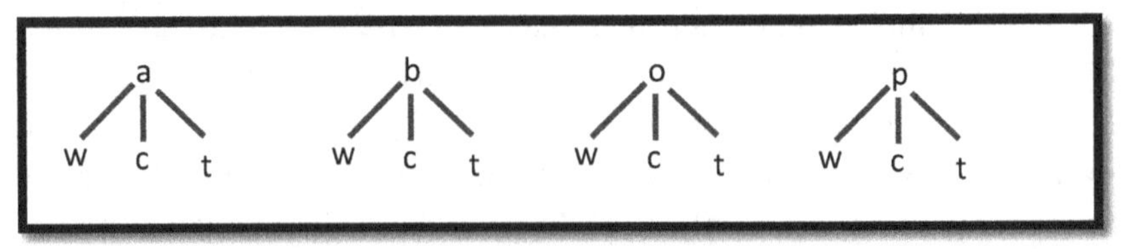

Fig. 11.2 A tree diagram representation of the cafeteria situation.

Another way of counting through a system, different from the Rule of Product, can be demonstrated through the following situation: *If John was invited to enroll in a teacher education program at five private and seven public universities, there are 5 + 7 choices of the program that he can make.* Indeed, John cannot be enrolled in more than one university and, therefore, the choices of the universities do not depend on each other. However, if one can concurrently enroll in a private and in a public university, there are 5 × 7 possibilities to enroll. Otherwise, the counting of possibilities is due to

The Rule of Sum. *If an object A can be selected in m ways and, independently from this selection, an object B can be selected in n ways, there are m + n ways to select either A or B.*

Consider now the following modification of the cafeteria situation. *A cafeteria offers three types of drinks – water (w), coffee (c), tea (t) – and four types of fruit – apples (a), bananas (b), oranges (o), and pears (p). Yet only banana can be offered when one buys coffee and only coffee; that is, no other drink allows for banana. How many ways can one buy a pair of drink and fruit?* This situation is described through the tree diagram of Fig. 11.3 and may be referred to as

The Rule of Sum of Products. *If an object A can be selected in m ways only k of which allow for n selections of an object B and the remaining m – k selections of A allow for r selections of B, then there are k × n + (m – k) × r ways to select the ordered pair (A, B).*

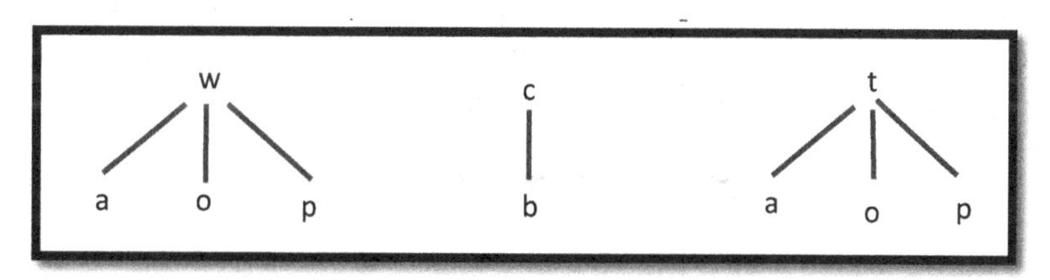

Fig. 11.3. A tree diagram representation of the modified cafeteria situation.

Remark 11.1. One can generalize the Rule of Product to the selection of *n* objects $A_1, A_2, \ldots, A_n$, with the selection choices

$m_1, m_2, \ldots, m_n$, respectively, so that the selection of the n-tuple ($A_1, A_2, \ldots, A_n$) can be done in $m_1 \times m_2 \times \ldots \times m_n$ ways.

Remark 11.2. If the selections of the objects $A_1, A_2, \ldots, A_n$ are not dependent on each other, then one can select either one of these objects in $m_1 + m_2 + \ldots + m_n$ ways.

Remark 11.3. The use of the Rule of Sum can be seen on a tree diagram through the development of the stems of the tree; for example, whatever a drink, each fruit can be selected in one way only and such selections are mutually independent. Therefore, four fruits can be selected in four ways (Fig. 11.1; four stems leading from drink to fruit). Likewise, three drinks can be selected in three ways (Fig. 11.2; three stems leading from fruit to drink). Consequently, one can rewrite the product 3×4 in the form $(1 + 1 + 1) \times (1 + 1 + 1 + 1)$ and the product 4×3 in the form $(1 + 1 + 1 + 1) \times (1 + 1 + 1)$.

11.2 From tree diagrams to permutations

Consider another real-life situation. *At a book sale, Anna, without looking, reaches into a box filled with books priced \$5 and \$10. If she selects two books, how much money should she be prepared to pay*? Using the Rule of Product, the following tree diagram can be created (Fig. 11.4) from which one can see that the sums \$10 and \$20 appear only one time and the sum \$15 appears twice. In other words, the appearance of \$15 twice is due to the irrelevance (in terms of the price Anna has to pay) of the order in which the differently priced books are selected. That is, a tree diagram shows different orders of objects selected.

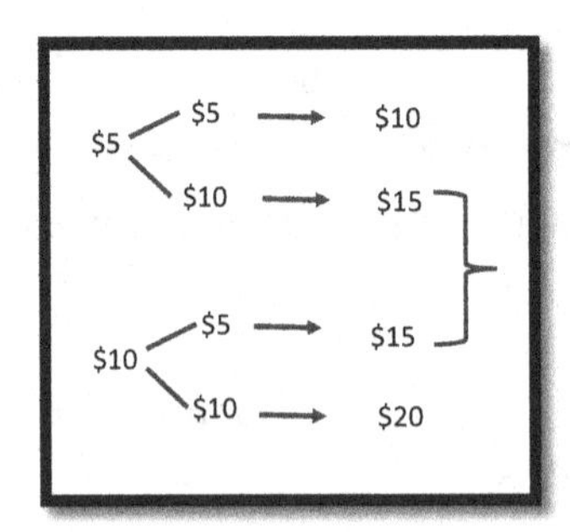

Fig. 11.4. The irrelevance of order in which two books are selected for the total price.

Consider a similar problem adopted from [National Council of Teachers of Mathematics, 2000, p. 52]. *I have pennies, dimes and nickels in my pocket, at least three coins of each denomination. If I took three coins out of the pocket, how much money could I have taken?* Fig. 11.5 shows a tree diagram solution to this problem. Whereas, according to the Rule of Product, there are $3 \times 3 \times 3 = 27$ sums of money that could be taken, only ten sums are different. This implies that a tree diagram approach based on the Rule of Product is not effective when the order of objects forming a path on a branch of the tree is immaterial. For example, the elements of the triple (1, 5, 10) can be summed up to 16 in six different orders. We saw this situation in Chapter 4, Section 4.2, when counting different orders in which three different size towers can form a triple of towers (see Fig. 4.3). Note that whereas the entire tree develops through the Rule of Product, the development of its branches at each step develops through the Rule of Sum. A tree diagram can also be used to demonstrate that there are six ways to form the sum 16 out of the three addends – 1, 5, and 10 – when each of the addends can be used one time only. As shown in Fig. 11.6, the first addend can be selected in three ways; yet, following this selection, the second addend can be selected in two ways only, thereby, leaving a single choice for the selection of the third addend. In other words, there are six ways to *permute* elements in the triple (1, 5, 10).

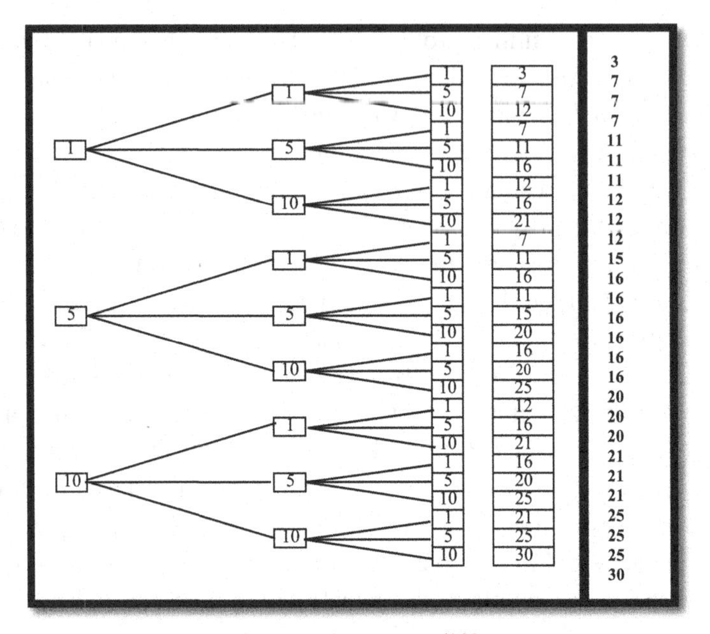

Fig. 11.5. Three coins – ten different sums.

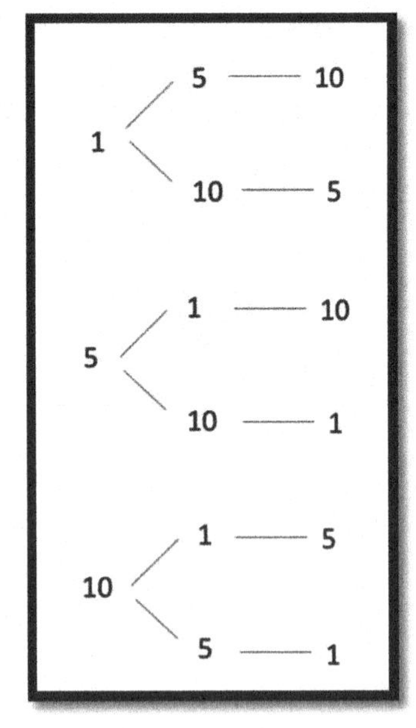

Fig. 11.6. The case of three (different) coins only.

11.3 Permutation of objects in the sets with distinct and repeated objects

The multiplicative structure of the tree diagram of Fig. 11.6 implies that when the (different) elements of the triple (1, 5, 10) may not be repeated, they can be permuted (i.e., arranged in different orders) in $3 \times 2 \times 1 = 6$ ways. Likewise, the (different) elements of the quadruple (1, 5, 10, 25) can be permuted in $4 \times 3 \times 2 \times 1 = 24$ ways. Indeed, for each permutation of the triple (1, 5, 10), the fourth element, 25, can be positioned in four ways around the elements 1, 5, and 10: to the right of 10, between 10 and 5, between 5 and 1, and to the left of 1. This leads to the product $4 \times 6 = 6 \times 4 = 1 \times 2 \times 3 \times 4 = 24$. In mathematics, the product of the first n natural numbers has the notation $n!$ (reads n factorial). That is, $1 \times 2 \times 3 = 3!$ and $1 \times 2 \times 3 \times 4 = 4!$. In general, according to the Rule of Product, in the case of n different symbols A_1, A_2, ..., A_n, the number of permutations of these symbols in the (artificial) word $A_1A_2...A_n$ is equal to $n!$

But what if we have a triple of integers two of which are the same, like (1, 1, 5)? The tree diagram of Fig. 11.7 shows that with two repeated integers among three, there are only three ways to permute them. Indeed, comparing the tree diagrams of Figs 11.5 and 11.6, one can see that when on the former figure, 10 is replaced by 1, each path on the tree, namely, 1 — 5 — 1, 1 — 1 — 5, and 5 — 1 — 1, is repeated twice. That is, as the notation 2! (= 2) represents the number of permutations of two objects, in the case of those two objects being the same, the number of permutations with three different objects has to be divided by 2!. Put another way, through the transition from the triple (1, 5, 10) to the triple (1, 1, 5), six permutations of the former triple are reduced by half. Alternatively, using the Rule of Sum of Products (Section 11.1), the first element of the triple can be selected in two ways, one of which, 5, allows for only one selection of the second element, and another one, 1, allows for two selections of the second element. In all three cases, there is only one way to select the third element; that is, there are $1 \times 1 \times 1 + 1 \times 2 \times 1 = 3$ ways to permute numbers in the triple (1, 1, 5).

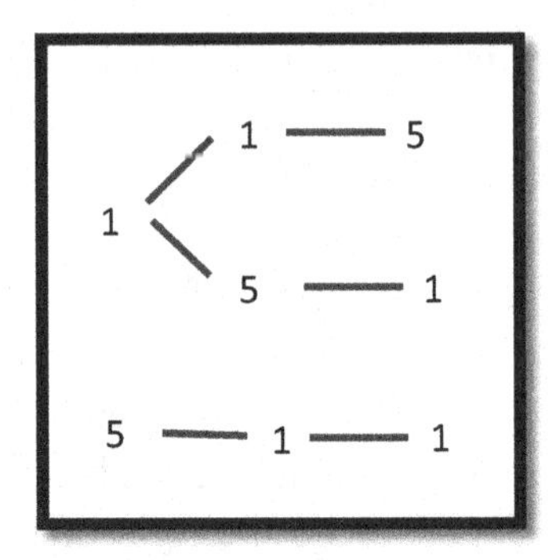

Fig. 11.7. Permutation of the elements of the triple (1, 1, 5).

How many permutations of letters are there in the word APPLE? If all five letters were different, the answer would be 5! (=120) permutations. With two repeating P's, the number of permutations of letters in the word APPLE should be smaller; let it be equal to N. For each of the N permutations, two P's can be permuted in 2! ways. By the Rule of Product, $2!\cdot N = 5!$, whence $N = \frac{5!}{2!} = 60$. Similarly, there are $\frac{5!}{2!} = 60$ ways to permute letters in the word GLASS. In the word SETTER the number of permutations of six letters with two E's and two T's is equal to $\frac{6!}{2!\cdot 2!} = 180$. In the word MISSISSIPPI with four I's, four S's, and two P's the number of permutations of the letters is $\frac{11!}{4!\cdot 4!\cdot 2!} = 34650$.

11.4 Using permutations in counting additive decompositions of integers

The counting technique developed in Section 11.3 can be applied to the towers in Figs 4.2 and 4.7 (Chapter 4). For example, the last (far-right) tower in Fig. 4.2 can be described through the "word" AAB with the number of permutations of its letters equal to $\frac{3!}{2!} = 3$. Likewise, the last tower in Fig. 4.7 can be described by the "word" AABB and the number of permutations of its four letters is equal to $\frac{4!}{2!\cdot 2!} = 6$.

In Chapter 4, a new strategy of decomposing an integer in two, three, and four addends with regard to their order by creating spaces between horizontally located blocks was introduced. Below, this strategy will be described in terms of permutations of letters in a word. To this end, consider the case of decomposing the number 10 in two addends with regard to their order (i.e., applying commutative property of addition to any pair of addends). Ten blocks representing the number 10 are shown in Fig. 11.8. These blocks can be put in two groups by selecting a space between two neighboring blocks. Therefore, the number of decompositions of ten in two addends arranged in all possible orders is equal to the number of ways such a space can be selected. There are nine spaces separating ten blocks[33]. That is, there are nine ordered decompositions of the number 10 in two (positive integer) addends. Formally, the situation can be represented through a nine-letter word $\underbrace{S\overline{S}\,\overline{S}\ldots\overline{S}}_{8\ letters}$ where the letters S and $\overline{S}$ stand, respectively, for the spaces selected (one space) and not selected (eight spaces). The number of permutations of letters in this word is equal to $\frac{9!}{8!}=\frac{8!\cdot 9}{8!}=9$. This not only confirms simple results of Chapters 4 and 6 (Sections 4.1 and 6.4, respectively) where nine ways of decomposing the number 10 in two ordered addends were listed, but the permutation of letters approach is used to confirm a more sophisticated way of finding the number of such decompositions. As the Common Core State Standards [2010, p. 6] put it, "students might rely on using concrete objects or pictures to help conceptualize and solve a problem ... check their answers to problems using a different method ... and identify correspondence between different approaches".

Fig. 11.8. A visual representation of the number 10.

---

[33] The number of spaces separating several lined-up blocks is always one smaller than the number of the blocks: one space separating two blocks, two spaces separating three blocks, three spaces separating four blocks, and, in general, $(n-1)$ spaces separating n blocks.

Likewise, decomposition of the number 10 in three addends arranged in all possible orders can be associated with a nine-letter word $\underbrace{SSS\,S\ldots S}_{7\,letters}$. Each permutation of the letters in this word represents a decomposition of 10 in three (differently ordered) addends. The total number of permutations of letters in the word with two letters S and seven letters $\overline{S}$ is equal to $\frac{9!}{2!\cdot 7!}=\frac{7!\cdot 8\cdot 9}{2\cdot 7!}=36$. This number, without using factorials, was already found in Chapter 4, Section 4.2.

To decompose the number 10 in four addends arranged in all possible orders, one has to find the number of permutations of letters in another nine-letter word $\underbrace{SSS\overline{S}\,\overline{S}\ldots\overline{S}}_{6\,letters}$. This number is equal to

$$\frac{9!}{3!\cdot 6!}=\frac{6!\cdot 7\cdot 8\cdot 9}{6\cdot 6!}=84.$$

In general, in order to decompose the number m in n addends arranged in all possible orders, $m \geq n$, note that there are $(m - 1)$ spaces separating m blocks and one has to select $(n - 1)$ spaces from $(m - 1)$ spaces. This leads to the word $\underbrace{SS\ldots S}_{(n-1)\,letters}\underbrace{\overline{S}\,\overline{S}\ldots\overline{S}}_{(m-n)\,letters}$ the number of permutations of letters in which is equal to $\frac{(m-1)!}{(n-1)!\cdot(m-n)!}$. In particular, when $m = 10$ and $n = 4$ the last fractional expression turns into $\frac{9!}{3!\cdot 6!}=\frac{6!\cdot 7\cdot 8\cdot 9}{6\cdot 6!}=84.$

11.5 Combinations without and with repetition

By selecting r objects from t objects, $t \geq r$, one creates what is called an r-combination of t objects. When all r objects are different, one talks about a combination without repetition (of objects). When among r objects there are several repeating objects, one talks about a combination with repetition (of objects). The task is to develop techniques (formulas) for counting combinations without and with repetition of objects. Fig. 11.9 shows a set of six different shapes out of which, as shown in Fig. 11.10, five different objects can be selected in six ways. That is, Fig. 11.10 shows that there exist six 5-combination without repetition of 6 objects. At the same time, if repetition of objects is allowed, one can select all five squares (or three squares and two circles, etc.) from the set of objects shown in Fig. 11.9.

To begin, note that each selection of 5 objects out of 6 objects without repetition can be represented through a 6-letter word with five letters Y and one letter N, where the rank of N in a word points to the rank of a shape in Fig. 11.9. So, the word YYYYYN, in which the rank of N is six, means that all shapes but the hexagon have been selected. Likewise, the word NYYYYY in which the rank of N is one, means that all shapes but the square have been selected. The number of permutations of letters in a word with five letters Y and one letter N is equal to $\frac{6!}{5!} = \frac{5! \cdot 6}{5!} = 6$. In that way, theory (i.e., a mathematical formula) confirms that there are 6 ways to select 5 objects out of 6 different objects; each way representing a 5-combination of 6 objects without repetition. In the general case of r-combination of n objects without repetition, the number of Y's is equal to r and the number of N's is equal to $n - r$. The number of permutations of letters in the word $\underbrace{YY\ldots Y}_{r\ letters}\underbrace{NN\ldots N}_{(n-r)\ letters}$ is equal to $\frac{n!}{r! \cdot (n-r)!}$. For example, because $\frac{8!}{5! \cdot (8-5)} = \frac{5! \cdot 6 \cdot 7 \cdot 8}{5! \cdot 3!} = 56$, there are 56 ways to check out five books ($r = 5$) out of eight different books ($n = 8$).

Fig. 11.9. A set of six different objects.

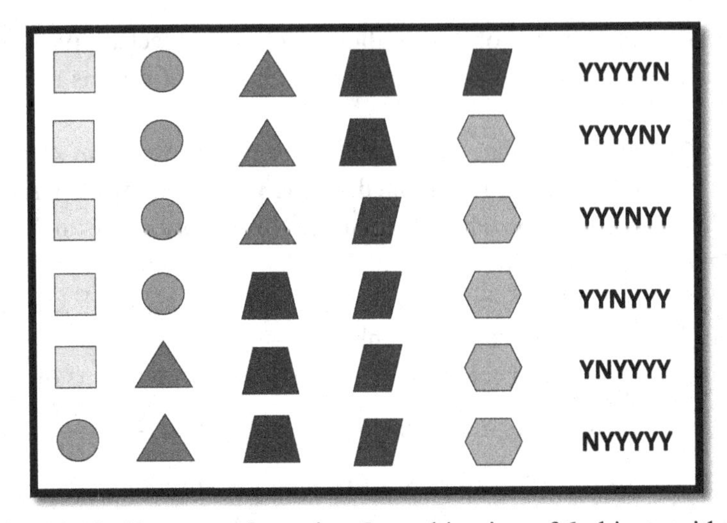

Fig. 11.10. Six ways of creating 5-combination of 6 objects without repetition.

Fig. 11.11 shows different ways of selecting two objects out of three different objects when selected objects may be identical (i.e., repeated). However, representation of a selection through a word with letters Y and N depends on whether the objects are repeated or not. In the case of selections with different objects the number of letters is equal to three (two Y's and one N), in the case of selections with repeated objects the number of letters is equal to four (two Y's and two N's). This demonstrates the need to modify representation of selections through a word to have the same number of letters regardless of selection. To this end, one has to avoid using letter N because, whereas the number of Y's stays the same indicating the number of objects selected, the number of N's varies. As shown in Fig. 11.11, what does not vary is the number of empty spaces between three objects. Using the letter *S* to stand for a separator between two shapes, a four-letter word with two letters Y and two letters S would accurately serve each selection regardless of whether the objects are repeated or not. So, in the case of the word YYSS the number of permutations of letters is equal to $\frac{4!}{2!\cdot 2!}=6$. In the general case of r-combination of n objects with repetition, the number of Y's is equal to r and the number of S's is equal to $n-1$. The number of permutations of letters in the word $\underbrace{YY\ldots Y}_{r\ letters}\underbrace{NN\ldots N}_{(n-r)\ letters}$ is equal to $\frac{(n-1+r)!}{r!\cdot(n-1)!}$. For example, there are 792 ways to buy five soft drinks (r = 5) out of eight different types (n = 8) of drinks as $\frac{(8-1+5)!}{5!\cdot(8-1)!}=\frac{12!}{5!\cdot 7!}=792$. Note that computations using factorials can be carried out using *Wolfram Alpha* as shown in Fig. 11.12.

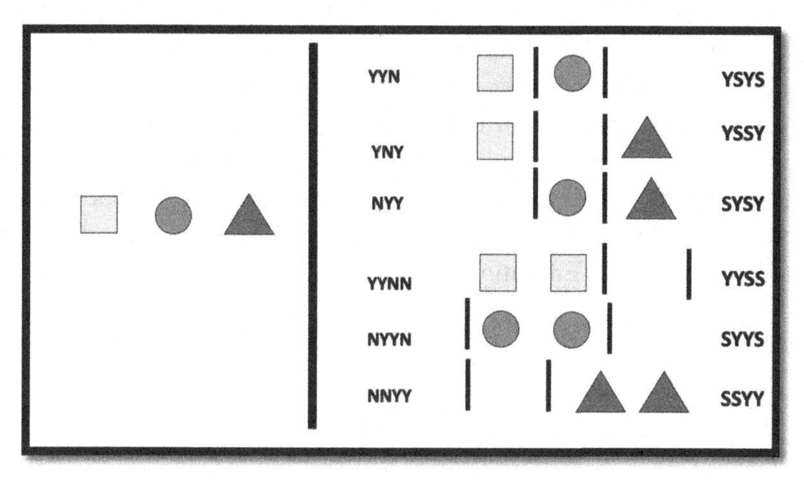

Fig. 11.11. There are six 2-combination of 3 objects with repetition.

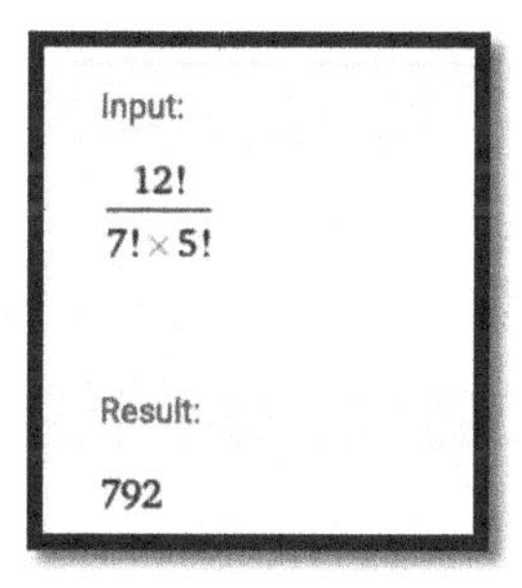

Fig. 11.12. Using *Wolfram Alpha* as a calculator.

Finally, one can use the technique of counting the number of permutations of letters in a word to answer the question that a second-grade student asked (see Chapter 4, Remark 4.1): *How many ways can one put five finger rings on three fingers?* Noting that three fingers (index, middle, and ring) are separated by two spaces, a 7-letter word SSYYYYY represents the case when all five finger rings are put on the ring finger (looking from left to right). The number of permutations of letters in this word, $\frac{7!}{\underbrace{2!}_{count\ S's} \cdot \underbrace{5!}_{count\ Y's}} = 21$, represents the number of ways five finger rings can be put on three fingers. This number was found without using factorials in Chapter 4, Remark 4.1 (see also [Abramovich, 2021]). An emphasis on the diversification of solution strategies in the United States expects students to "check their answers to problems using a different method ... understand the approaches of others ... and identify correspondence between different approaches" [Common Core State Standards, 2010, p. 6]. According to mathematics educators in Singapore, students' ability to "devise different strategies to solve an open-ended problem ... [belongs to] key competencies that are important in the 21st century" [Ministry of Education Singapore, 2020, p. 12].

CHAPTER 12: ELEMENTS OF PROBABILITY THEORY AND DATA MANAGEMENT

12.1 Teaching about chances

Teaching basic ideas of probability theory begins at the early elementary level by asking questions about chances of something to happen. What are the chances of rain (or snow) on the first day of a school year? What are the chances of snow (or rain) on the last day of a school year? Whereas an answer to each question depends on whether teaching happens in Northern of Southern Hemisphere, it is important for a child to understand that in the above two cases the chances are quite different. In a non-weather context, what are the chances to pick up (without looking) a penny from a box with three pennies and one dime? Here, it is important for a child to understand that without looking an outcome is uncertain; otherwise, an outcome is most likely certain. Still, within this uncertainty of an experimental situation, there should be more chances for a penny (just as there are more chances for rain in September than for snow in June in the Northern Hemisphere). But just as it may be snowing in June, a child might pick up the dime. Nonetheless, just as snow is a rare occurrence in a 100-year span, picking up the dime in a course of 100 trials would not happen often. Through a conversation, a child should appreciate the experimental character of situations associated with uncertain outcomes.

Teaching about chances can be supported by the appropriate use of physical and digital teaching tools. At the upper elementary level, students "use proportionality and a basic understanding of probability to make and test conjectures about the results of experiments and simulations" [National Council of Teachers of Mathematics, 2000, p. 248] and "learn about the importance of representative samples for drawing inferences" [Common Core State Standards, 2010, p. 46].

In the study of probability, experiments with equally likely outcomes are of a special importance. For example, tossing a fair coin and rolling an unbiased die can be set as experiments with equally likely outcomes. It is equally likely to have either head or tail when tossing a coin and either one or six when rolling a die. An Italian mathematician Gerolamo Cardano (1501–1576) is credited with the definition of the concept of probability as the ratio (Chapter 9) of the number of favorable outcomes to the total number of *equally likely* outcomes within a certain experimental situation. This classic definition is commonly accepted nowadays in school mathematics. Whereas saying that there are three chances out of four to pick up a penny in the above situation can be seen as an intuitive conclusion, comparing chances of two situations requires numerical measurement of chances. When solving probability problems with equally likely random outcomes, this definition gives an applied flavor to fractions in the range [0, 1] as numerical characteristics of what is considered probable and makes it possible to compare chances by using proper fractions (or their decimal equivalents). Thus, according to Cardano, the chances (alternatively, likelihood or probability) to pick up a penny is the fraction 3/4. But chances (probabilities) related to different situations, in order to be compared, have to be computed (measured) first. Mathematical actions of that kind should not be considered through the simplistic lens, like measuring perimeters and areas of basic geometric shapes through selecting an appropriate measurement tool of the modern-day geometry software. For example, whereas chances to select a penny from the boxes with three pennies and one dime, and five pennies and two dimes, respectively, are close, the chances for the former box are higher as $3/4 > 5/7$. And until one knows how to express chances through a common fraction and how to compare fractions, the comparison of chances can only be done intuitively or through an experiment.

When tossing a coin and knowing with certainty that it would not fly, one, however, cannot predict how exactly it would fall. But through tossing a coin many times and recording the results, one can recognize how randomness turns into regularity. Indeed, one can check to see that out of 100 tosses, a coin would turn up head or tail somewhere in the range [40, 60]; that is, an *experimental* probability of head or tail, computed as the ratio of the number of times a desired outcome occurred to the total number of trials, would be a number in the range [0.4, 0.6].

	A	B	C	D	E	F	G	H
1							596	
2						0.0001	0.0596	
3		1	1	1	1	HHHH		
4		1	0	0	1			
5		0	1	1	1			
6		0	0	0	1			
7		0	1	0	1			
8		0	1	1	0			
9		0	1	1	1			
10		1	1	0	0			
11		1	1	0	0			
12		0	0	1	1			
9998		1	1	1	1	HHHH		
9999		0	1	1	0			
10000		1	0	1	0	HTHT		
10001		0	1	0	1			
10002		1	0	0	0			

Fig. 12.1. Tossing four coins.

A more sophisticated experimentalist can attempt to measure the results of possible outcomes when a coin is tossed, say, four times, in a large series of experiments. A surprising result, which first can be established experimentally by using a spreadsheet, is that having four heads (HHHH) has the same probability as having first head, then tail, and then again head and tail (HTHT). The spreadsheet of Fig. 12.1 shows that as a result of 10,000 (simulated) tosses of a coin the incident HHHH happened about as many times as the incident HTHT (616 vs. 621); in other words, the chances for HHHH and for HTHT are measured, respectively, by two fractions (616/10000 = 0.0616 and 621/10000 = 0.0621) differing by 5/10000 = 0.0005 (Fig. 12.1, cell F2). Multiple repetitions of this computational experiment would demonstrate that both incidents have almost equal chances within a *large* series of experiments. The programming of the spreadsheet of Fig. 12.1 is included in Chapter 13, Section 13.12.

12.2 Randomness and sample space

In this section, several basic concepts associated with the probability strand will be explained. The first one is randomness – a characterization of the result of an experiment that is not possible to predict but, nonetheless, is often possible to measure on the scale from zero to one, with zero assigned to something impossible (e.g., a coin flies when tossed in the still air) and one assigned to something certain (e.g., a coin falls when tossed in the still air). One can say that randomness does not lead to a credible pattern, although it is often difficult to conclude whether there is no pattern in a sequence of events. Nonetheless, thinking about randomness in a very informal way, can help one understand the "difference [between] predicting individual events and predicting patterns of events" [Conference Board of the Mathematical Sciences, 2001, p. 23]. In the case of tossing a coin, whereas one cannot predict how exactly it would fall, it is not impossible to have head and tail alternating in, say, five tosses, or even to have five heads or tails in a row. Yet, due to experience, one can predict a pattern in the sequence of 100 tosses. Whatever a prediction, the question to be answered is how to measure the likelihood (chances) of the prediction.

The need for measuring chances leads to the concept of the sample space of an experiment with a random outcome that is defined as the set of all possible incidents associated with this experiment. A simple example is the sample space of rolling a six-sided die (with the number of spots on the sides ranging from one to six) comprised of six outcomes: {1, 2, 3, 4, 5, 6}. Assuming that we deal with a fair (unbiased) die, all outcomes may be considered equally likely. Under this assumption, one can say that there is one chance out of six to cast any of the six numbers (spots).

N\D	*0*	*1*	*2*
0	25	15	5
1	20	10	0
2	15	5	
3	10	0	
4	5		
5	0		

Fig. 12.2. A sample space of changing a quarter into pennies, nickels and dimes.

A more complicated example is the sample space of an experiment of changing a quarter into pennies, nickels and dimes. The chart of Fig. 12.2 shows a twelve-element sample space, where numbers in the top row and the far-left column show the ranges for dimes and nickels, respectively, and the remaining numbers show the corresponding number of pennies in a change. For example, the triple (5, 2, 1) stands for five pennies, two nickels, and one dime, so that $5 \cdot 1 + 2 \cdot 5 + 1 \cdot 10 = 25$. Alternatively, the sample space can be described as the following set of twelve triples of numbers {(25, 0, 0), (20, 1, 0), (15, 2, 0), (10, 3, 0), (5, 4, 0), (0, 5, 0), (15, 0, 1), (10, 1, 1), (5, 2, 1), (0, 3, 1), (5, 0, 2), (0, 1, 2)}, where the elements of each triple describe, respectively, the number of pennies, nickels and dimes in a change. Note that there is no reason to assume that it is equally likely to get any combination of the coins from a change-making device and Fig. 12.2 represents a sample space with not equally likely outcomes. Therefore, given a change-making device, it is only experimentally that one can determine chances for a specific combination of coins in a change. Alternatively, for the purpose of problem solving, one can assume that it is equally likely to get any combination of the coins in a change out of the total twelve. Under this assumption, one can say that the probability of not having pennies in the change is equal to 1/4 (as 3/12 = 1/4).

12.3 Different representations of a sample space and Pascal's triangle

A sample space of an experiment with random outcomes may have different representations. Here is an example: the sample space of tossing two coins can be represented in the form of a table and a tree diagram as shown in Fig. 12.3 and Fig. 12.4, respectively. Note that the outcomes of the two tosses are independent; that is, the outcome of the second toss does not depend on what happened on the first toss. This independence is reflected in the very form of the tree diagram: each of the two possible outcomes of the first toss affords the same two outcomes for the second toss. This, however, is not always the case and in more complex situations drawing a tree diagram to represent a sample space may be a difficult proposition.

An outcome of an experiment is an element of its sample space. For example, the sample space of the experiment of tossing two coins shown in Fig. 12.3 (or Fig. 12.4) consists of four outcomes. Further, outcomes may be combined to form an event. For example, the outcomes HH and TT form an event that both coins turn up the same. Depending on specific conditions, the outcomes of an experiment may or may not be equally likely. All the outcomes of the experiments of tossing coins or rolling dice described in the above examples are considered equally likely, assuming that one deals with fair coins of unbiased dice. An assumption about equally likely outcomes is a theoretical assumption – after all, when we deal with a coin (or a die), we *assume* that it is a fair coin (or an unbiased die).

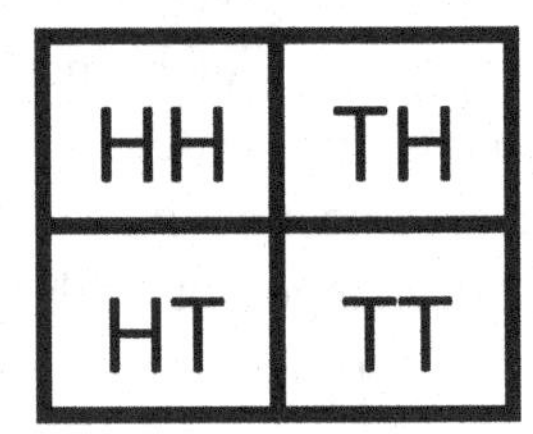

Fig. 12.3. The sample space of tossing two coins in the form of a table.

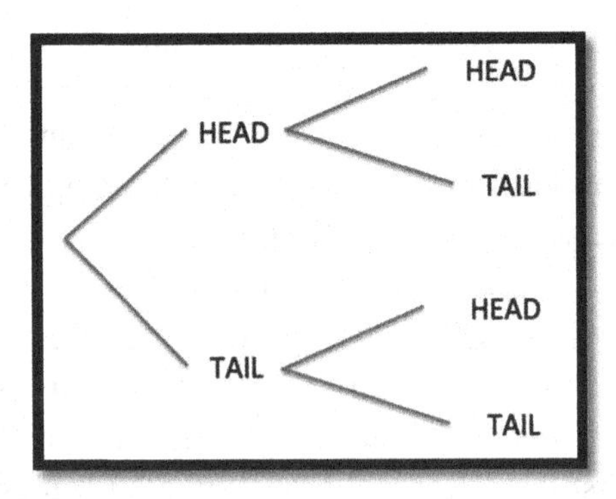

Fig. 12.4. The sample space of tossing two coins in the form of a tree diagram.

Due to two possible outcomes resulted from tossing a coin, the following special cases can be recorded. When $n = 1$ (tossing one coin), the sample space $\Omega_1 = \{H, T\}$ can be associated with the pair of integers (1, 1) meaning that the set Ω_1 comprises one outcome with one head and one outcome with no head. When $n = 2$ (tossing two coins), the sample space $\Omega_2 = \{HH, HT, TH, TT\}$ can be associated with the triple of integers (1, 2, 1) meaning that the set Ω_2 comprises one outcome with two heads, two outcomes with one head, and one outcome with no head. When $n = 3$ (tossing three coins), the sample space $\Omega_3 = \{HHH, HHT, HTH, THH, HTT, THT, TTH, TTT\}$ can be described through the quadruple of integers (1, 3, 3, 1) meaning that the set Ω_3 comprises one outcome with three heads, three outcomes with two heads, three outcomes with one head, and one outcome with no head. One can see that the above three strings of numbers – (1, 1), (1, 2, 1), (1, 3, 3, 1) – are the second, third, and fourth rows, respectively, of the famous numeric structure (Fig. 2.5) known as Pascal's triangle (see also Chapter 5, Section 5.2, Remark 5.3). According to Kline [1985], Pascal came across this triangle through recording sample spaces of experiments of tossing coins from where one can determine chances of having a certain result of an experiment. Indeed, the sample space $\Omega_1 = \{H, T\}$ suggests there is one chance out of two to have exactly one head when tossing one coin, the sample space $\Omega_2 = \{HH, HT, TH, TT\}$ suggests there are two chances out of four to have both head and tail when tossing two coins, and the sample space $\Omega_3 = \{HHH, HHT, HTH, THH, HTT, THT, TTH, TTT\}$ suggests there are three chances out of eight to have exactly two heads when tossing three coins.

Also, comparing two consecutive rows of Pascal's triangle, one can see that in each row beginning from the third, any number greater than one is the sum of the two numbers immediately above it. For example, the number 2 in the third row is the sum 1 + 1, both numbers 3 in the fourth row are the sums 1 + 2, and so on. This observation about the arithmetic property of Pascal's triangle is consistent with the following comparison of $\Omega_1 = \{H, T\}$ and $\Omega_2 = \{HH, HT, TH, TT\}$: the pair (*HT*, *TH*) representing a single event within Ω_2 develops from two different outcomes, *H* and *T*, within Ω_1 by attaching *T* to *H* and *H* to *T* followed by adding the two modifications of *H* and *T*. Likewise, the triple (*HHT*, *HTH*, *THH*) representing a single event within Ω_3 develops from two different events, *HH* and *HT-TH*, within Ω_2 by attaching *T* to *HH* and H to both *HT-TH* followed by adding the two modifications of *HH* and *HT-TH*.

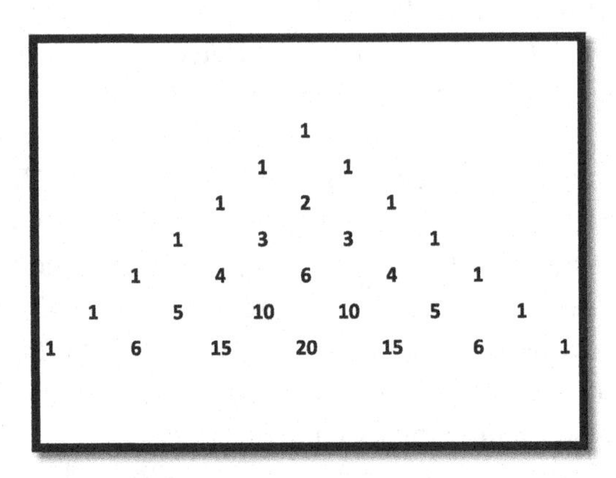

Fig. 12.5. Pascal's triangle.

Outcomes are not always equally likely. Indeed, when having a holder with three red and two black markers, it is not equally likely (without looking) to pick up either a red or a black marker. Likewise, events may be independent and dependent. In the case when from the first draw of a marker from the holder, the marker is returned to the holder, the outcome of picking up a red marker on the second draw does not depend on which color marker was picked up on the first draw. At the same time, if the marker is not returned, the outcome of picking up a red marker on the second draw depends on what happened on the first draw.

The sample space of rolling two dice and recording the total number of spots on two faces can be represented in the form of an addition table (Fig. 12.6) from where it follows that the highest chances are for having seven as the total number of spots on two dice. Here, one can connect probability concepts to the concept of integer partitions into like addends (see Chapter 4). Indeed, whereas twelve has more partitions than seven, in the context of rolling two dice partitions are limited to the addends that are not greater than six. Under such condition, the closer a number (different from six) to six is, the more ordered partitions into two integer addends exist. For example, the sum 12 has only one possibility, $12 = 6 + 6$, and the sum 7 has six possibilities, $7 = 1 + 6 = 6 + 1 = 2 + 5 = 5 + 2 = 3 + 4 = 4 + 3$. At the same time, the sum 6 has only five possibilities.

	1	2	3	4	5	6
1	2	3	4	5	6	7
2	3	4	5	6	7	8
3	4	5	6	7	8	9
4	5	6	7	8	9	10
5	6	7	8	9	10	11
6	7	8	9	10	11	12

Fig. 12.6. The sample space of rolling two dice.

12.4 Understanding chances in a computer environment

The following two problems aim at elementary school students' learning to compare chances before they are taught fractions which serve as tools for computing and comparing chances. For example, mathematics educators in Chile focus their attention on "using playful and everyday situations to illustrate how to quantify chance" (Felmer et al., 2014, p. 213). In a more general context, South African mathematics educators advised that "play-based learning should be promoted because it is important component of active learning of mathematics" [Department of Basic Education, 2018, p. 20]. In particular, following these pedagogic experiences, students can learn to compare and quantify chances visually which is different from the use of intuition or conducting an experiment even within an everyday situation.

Problem 12.1. *Billy wants to eat red M&Ms only. There are two bags of M&Ms, plain and peanut, available. If there are 4 red M&Ms out of 10 total in the first bag and 10 red M&Ms out of 20 total in the second bag, for which bag does Billy have more chances to get a red M&M? What are these chances? Use sliders to create the bags of M&Ms within the spreadsheet of Fig. 12.7. You may compare the height of the bar graphs to answer the first question.*

Problem 12.2. *Billy ate (by using sliders) one red M&M from each bag (Fig. 12.8). How many red peanut M&Ms does Billy have to eat now in order to give Mary more chances to get a red plain M&M than a red peanut M&M? Use sliders to create new bags. What are Mary's chances for getting a red M&M for each bag? Once again, bar graphs represent chances.*

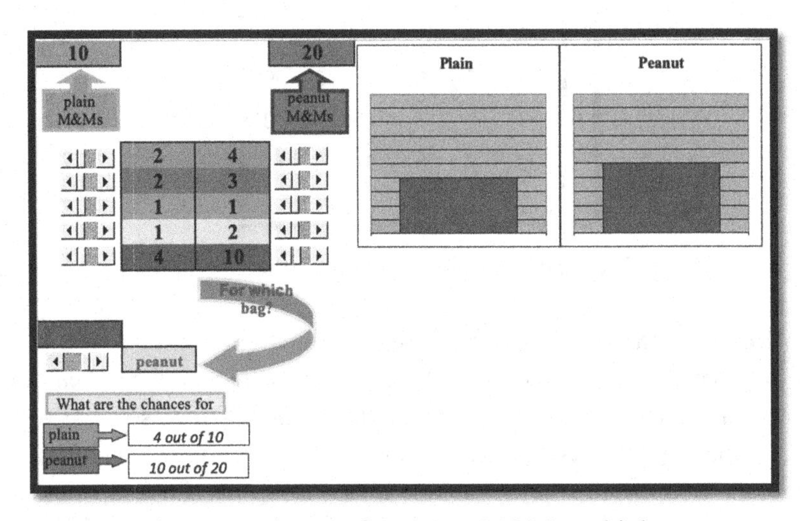

Fig. 12.7. Chances for peanut M&M are higher.

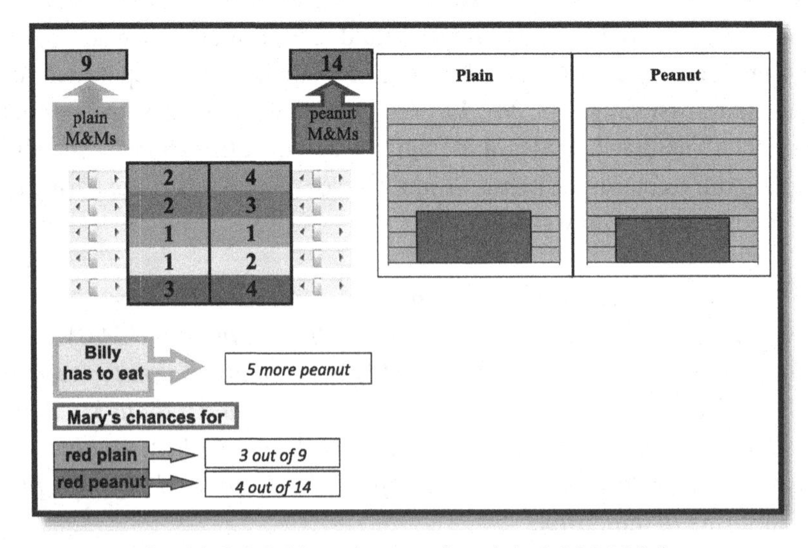

Fig. 12.8. Making chances for plain M&M higher.

From the quantitative perspective, Problem 12.1 requires the comparison of two fractions, 4/10 and 10/20. When students do not know fractions and, moreover, do not know how to compare fractions, they have no idea for which bag the chances to pick up a red M&M are higher. An intuitive grasp of fractions as quantities can be developed using their iconic representations. To this end, spreadsheet-based bar graphs can be employed for an intuitive, visual comparison of fractions. Such graphs are the analogues of those used to represent and compare whole numbers (see Chapter 1, Section 1.2, Fig. 1.10). In such a way, students can acquire an informal, intuitive understanding of quantitative relationships between chances before they actually learn about formal ways of expressing and comparing chances using fractions. In other words, the comparison of chances can be developed first by using mathematical visualization in the iconic environment of a spreadsheet. The use of fractions as symbolic representations of the chances develops later, on the foundation provided by visualization. That is, students can build their leaning of formal operations with fractions on the knowledge they received in a fun and informal, yet mathematically meaningful, way. Figs 12.7 and 12.8 present students' (correct) responses to questions posed in Problems 12.1 and 12.2, respectively. For more details on this project see [Abramovich, Stanton, and Baer, 2002]. The programming of the spreadsheets of Figs 12.7 and 12.8 is included in Chapter 13, Section 13.14.

12.5 Fractions as tools in measuring chances

Application of arithmetic to geometry brings about the idea of using a numeric measure of the chances (probability) of an event. One of the ways to introduce the concept of fraction as an applied tool is through using fractions as a measure of the chances (probability) of an event. The idea of a geometric representation of a fraction enables its use as a means of finding the probability of an event (alternatively, measuring its chances).

Consider equally likely outcomes of the experiments with tossing a coin and rolling a die. The meaning of the words *equally likely* can be given a geometric interpretation through representing such outcomes as equal parts of the same whole. In the case of tossing two coins, we have four equally likely outcomes, which can be represented in the form of a rectangle divided into four equal parts (Fig. 12.3) so that each part is represented by a unit fraction 1/4. It is this fraction that can be considered as the value of the likelihood (probability) of any of the four (equally likely) outcomes. Dividing the rectangle in four equal parts as shown in Fig. 12.3 demonstrates two things: (i) a table representation of the sample space of tossing two coins; (ii) according to the area model for fractions (Chapter 7), each of the four parts represents the product (1/2)(1/2), where each factor is the probability of head/tail for each of the two tosses. Extending the number of tosses beyond two, one can measure the probability of alternating heads and tails in, say, five tosses by the number $\left(\frac{1}{2}\right)^5 = \underbrace{\frac{1}{2} \cdot \ldots \cdot \frac{1}{2}}_{5\ times}$. One can see that the same number also measures the probability of having five heads (or tails) in a row.

In the case of rolling two dice, each of the 36 outcomes that comprise the table-type sample space (Fig. 12.4) is equally likely and, therefore, the fraction 1/36 is the measure of each outcome. At the same time, the likelihood of the event that after rolling two dice, the result is either eight or nine spots on both faces is measured by the fraction of the table filled with either eight or nine. As the two numbers appear nine times in the (addition) table, the probability of this event is 9/36 or (after reduction to the simplest form) 1/4. In that way, one can conclude that the chances of having HH after tossing two coins are the same as having the sum of either eight or nine on both faces when rolling two dice. Apparently, without using fractions as measuring tools for chances (likelihood, probability) this conclusion would not be possible.

Consider the case of rolling three dice. In order to find the total number of outcomes of this experiment, the Rule of Product (Chapter 11, Section 11.1) can be used and represented through the tree diagram of Fig. 12.9. Because each die has six outcomes independent of the outcomes of other two dice, the total number of outcomes is the product $6 \times 6 \times 6 = 216$. In order to roll 10 on three faces, one has to find the number of ordered partitions of the number 10 in three positive integer addends not greater than six. As was shown in Chapter 4, Section 4.2, there are eight triples – (1, 1, 8), (2, 2, 6), (2, 4, 4), (3, 3, 4), (1, 2, 7), (1, 3, 6), (1, 4, 5), (2, 3, 5) – the sums of elements of which are equal to the number 10 and the elements in each triple can be permuted, respectively, in 3, 3, 3, 3, 6,

6, 6, 6 ways. The sum of the last eight numbers is 36. Because the triples (1, 1, 8) and (1, 2, 7) include elements greater than six, their apparent exclusion from consideration (in the context of rolling six-sided dice) implies that the numbers 3 and 6 have to be subtracted from 36. Therefore, there are 27 ways to cast 10 when rolling three dice. Consequently, the probability of such event is the ratio 27/216 = 1/8 = 0.125. This theoretical result can be confirmed computationally by rolling three dice, say, 5000 times. Such a spreadsheet, the programming of which is included in Chapter 13, Section 13.13, is shown in Fig. 12.10. By connecting additive decompositions of whole numbers to rolling dice, elementary teacher candidates and their future students can appreciate one of the major ideas that relate mathematics to real life: mathematical models have to be consistent with practical applications. Such connectivity of theory and practice is in line with many features of STEM education.

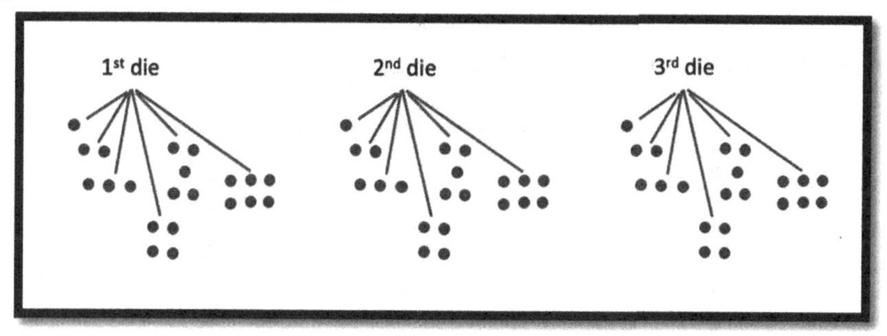

Fig. 12.9. A tree diagram of rolling three dice.

	A	B	C	D	E
1	Die 1	Die 2	Die 3		0.1246
2	5	4	4	0	
3	4	2	6	0	
4	1	3	3	0	
5	2	5	5	0	
6	3	5	4	0	
7	6	2	3	0	
4999	1	6	4	0	
5000	3	2	5	1	
5001	1	5	2	0	

Fig. 12.10. Experimental probability to cast 10 when rolling three dice 5000 times is 0.1246 (cell E1).

12.6 Graphic representations of numeric data

Suppose a die was rolled 20 times and the number of spots on the die assumed the following twenty values 1, 6, 4, 1, 5, 6, 1, 2, 3, 3, 5, 6, 6, 2, 3, 5, 6, 4, 4, 6. These values can be written in the non-decreasing order 1, 1, 1, 2, 2, 3, 3, 3, 4, 4, 4, 5, 5, 5, 6, 6, 6, 6, 6, 6, and then organized in different graphical formats. One such format is a graph called a line plot (Fig. 12.11). Reading the graph from left to right, one can see that the first three dots indicate the number of occurrences of 1's, the next two dots – the 2's, …, and, finally, the last six dots – the number of occurrences of 6's. Another graphic representation is called a histogram (Fig. 12.12). Reading the histogram, one can recognize that the height of each of the six bars indicate the number of occurrences of each of the six spots on a face of the die. As mentioned in [Association of Mathematics Teacher Educators, 2017, p. 52], "Well-prepared beginners of mathematics in Pre-K to Grade 2 understand that the foundations of statistical reasoning begin with collecting and organizing data".

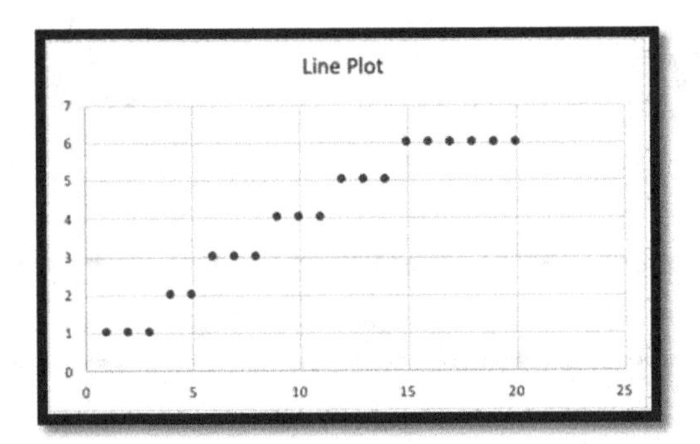

Fig. 12.11. A line plot constructed with *Excel*.

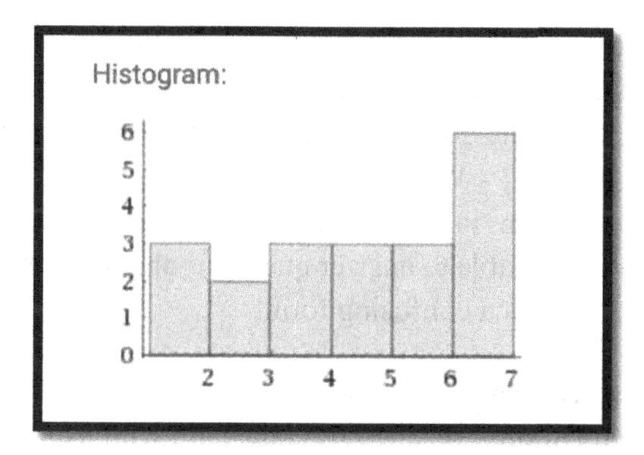

Fig. 12.12. A histogram constructed with *Wolfram Alpha*.

0	9	5			
1	0				
2	4	2			
3	5	7			
4	4	5			
5	8	7	3		
6	0				
8	1	1	4	9	9
9	4	5			

Fig. 12.13. Stem and leaf plot.

Numeric data can also be presented in the stem and leaf plot format. For the set of 20 one-digit numbers the stem may be defined as one of the numbers and the leaf may be defined as the number of occurrences of that number in the set. However, more common is to have one number representing the stem and having several leaves associated with this stem, something that a set of one-digit numbers does not follow. For example, in plotting the set of twenty integers 58, 35, 60, 57, 44, 53, 94, 81, 81, 24, 10, 95, 9, 45, 37, 5, 84, 89, 22, 89, the stem of the plot may be defined as the tens' digits and leaves of the plot are the ones' digits. The corresponding stem and leaf plot is shown in Fig. 12.13. In the presence of three-digit numbers, a stem may be the face value for the largest place value and the leaves are two-digit numbers. For example, the line 3| 0 7 45 89 may be read as the set {3, 37, 345, 389}. The development of a skill of correct interpretation of statistical graphs "is a desirable outcome of the Singapore mathematics curriculum" [Wu and Wong, 2009, p. 232]. While this outcome relates to secondary mathematics curriculum in Singapore, future teachers of the primary school pupils should be able to answer questions about graphs which often appear on the Internet in a confusing form.

12.7 Measures of central tendency

Suppose that five children decided to contribute with as much money as each of them had towards a common goal of buying a new basketball. Their individual contributions were $3, $3, $5, $8, and $11 totaling $30. Also, they wanted to know how could $30 be collected from equal contributions and divided the number 30 by 5 to get 6. Put another way, $6=\frac{3+3+5+8+11}{2}$. Statistically speaking, the number 6 can be used as a characteristic of their contributions; it is called the mean (or average) of the numbers 3, 3, 5, 8, and 11. Note that none of the five numbers is equal to six. Yet, the number 6 can be used to describe the five numbers in terms of fair sharing of moneys. In general, given the set of n numbers $x_1, x_2, \ldots, x_n$, the mean value, $\overline{x}$, is defined as $\overline{x}=\frac{x_1+x_2+\ldots+x_n}{n}$.

One can also look at the above five contributions to see whether some children from the group contributed the same amount into the purchase of a basketball. An observation can be made that two children each contributed $3, something that may be considered the most common contribution among the five. In statistics, this number is called the mode. In general, if in the set of numbers there exist a single number or several numbers that occur most frequently, each one is called the mode. That is, the set of numbers may have no mode or may have one or several modes. Often, numbers collected for statistical evaluation have to be arranged in the increasing (or non-decreasing) order. In such case, depending on the total of the numbers collected, there is either a number in the middle of the list or there are two neighboring numbers equidistant from the first and the last numbers on the list. Either the number in the middle or the average of the two numbers in the middle of the list is called the median. For example, for the list 3, 3, 5, 8, 11, the median is 5, and for the list 2, 3, 4, 6, 7, 8, the median is 5 as well. That is, like the mean, the median may or may not be a part of the list. In statistics, the mean, median, and mode are considered the most basic characteristics through which a data set can be described, and they are called the measures of central tendency. These measures might be good characteristics to be used in one case and weak characteristics to be used in another case.

Indeed, whereas the set of six numbers {2, 3, 4, 6, 7, 8} does not include their mean and median (both are equal to 5), the number 5 does not deviate much from either the smallest or the greatest. At the same time, the set of six numbers {0, 1, 2, 2, 2, 23} has the same mean, 5, as the former set; yet the number 5 deviates significantly from its largest number. This observation implies that measures of central tendency may not be sufficiently good characteristics of a set of numbers in terms of their (the numbers') deviation from the three characteristics. Therefore, new tools of statistical analysis have to be developed. Such tools are beyond the scope of this textbook.

CHAPTER 13: DEVELOPING TECHNOLOGY USED IN THE BOOK

13.1 Constructing fraction circles using the *Geometer's Sketchpad* (GSP)

A fraction circle is a physical representation of a unit fraction. As 1/2 is the largest unit fraction, the largest fraction circle is one-half. How can one construct an image of the one-half using GSP? Recall that a fraction circle represents a sector of a circle defined by the center from which two radii, forming an angle enclosed by an arc, stem. Thereby, three parameters have to be considered: the location of the center, the length of the radius, and the angle formed by two radii (alternatively, degree measure of the arc). The following seven steps will be presented with reference to Fig. 13.1. Step 1 is to construct a segment AB the left endpoint (A) of which is the center of a circle the half of which is the object to be constructed. This segment represents a radius of the circle. The next four steps deal with the construction of a diameter. To this end, one can either rotate the right endpoint (B) of the segment about its left endpoint (A) by 180° or reflect the right endpoint (B) in the center (A). The rotation requires one to designate the left endpoint (A) as the center of rotation: double click at point A (step 2) and then click (or highlight) the right endpoint (B) (step 3). To rotate, one has to open the *Transform* menu, choose *Rotate* and enter into the resulting dialogue box the number 180 as the measure of the angle of rotation. Clicking at the button *Rotate* yields (step 4) a point (C) to the left of the center (A). Connecting the so constructed point (C) with the center (A) completes (step 5) the construction of the diameter CB.

The next two steps deal with the construction of an arc supported by the diameter. To this end, one has to highlight three points – (A), (B), (C) – in that order, then open the *Construct* menu and choose *Arc on circle.* As a result, the fraction circle is constructed (step 6). To fill it with color, after clicking at the arc, one opens the *Construct* menu and in the line *Arc interior* selects either *Arc segment* or *Arc sector* (step 7).

Another way of constructing the third point (C) is to highlight the right and left endpoints (B) and (A) – in that order, and in the *Transform* menu to choose *Mark vector.* Then highlight point (A), in the *Transform* menu choose *Translate*, and in the resulting dialogue box hit the button *Translate*. Now, one has to repeat steps used after the third point (C) was constructed. Finally, one can construct a segment (making sure it is highlighted) and in the *Construct* menu choose *Midpoint*. In that way, the midpoint would play the role of the center of the arc (circle); to construct the arc one has to highlight three points on the diameter with the center being highlighted first and then in the *Construct* menu choose *Arc on circle*. Finally, by clicking at the arc, one opens the *Construct* menu and in the line *Arc interior* selects either *Arc segment* or *Arc sector*.

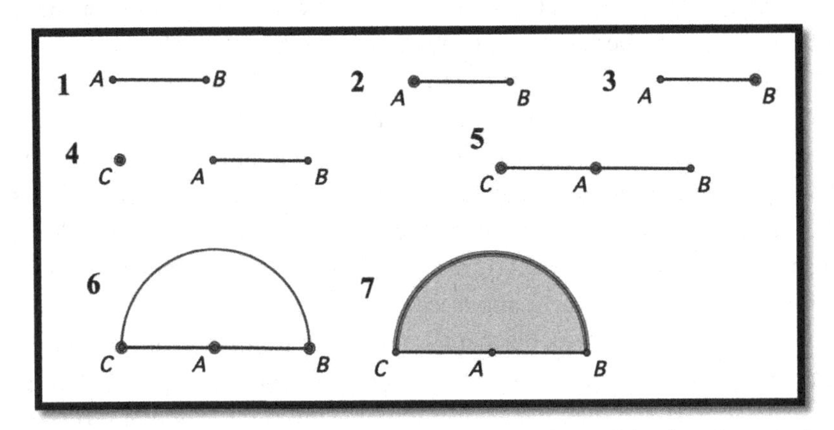

Fig. 13.1. Seven steps leading to the construction of the fraction circle one-half.

The next largest unit fraction is 1/3. In order to construct the fraction circle one-third, the first three steps shown in Fig. 13.1 have to be completed and, as the fourth step, point (B) has to be rotated about point (A) by 120° (=360°/3). The rotation yields point (C) (Fig. 13.2). The rest repeats steps 5 – 7 described in the construction of the fraction circle one-half. Similarly, other fraction circles can be constructed keeping in mind that the angle of rotation in the construction of the fraction circle $1/n$ is $360°/n$.

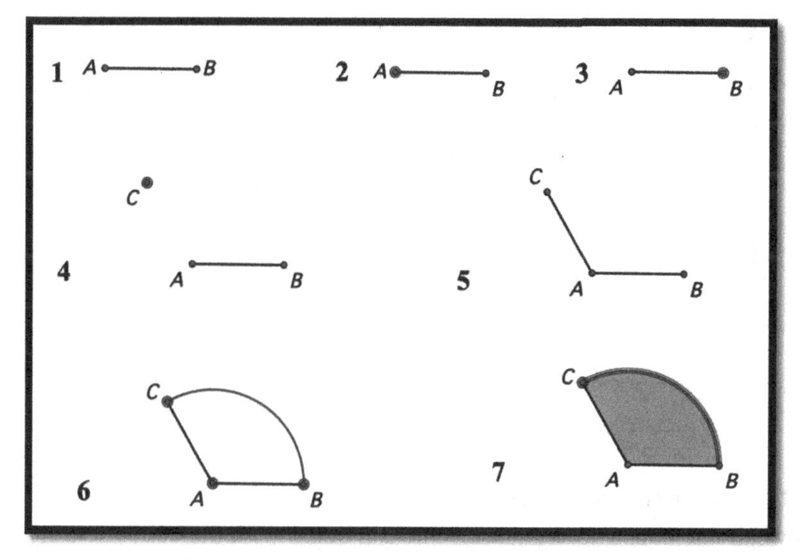

Fig. 13.2. Seven steps leading to the construction of the fraction circle one-third.

13.2 Constructing an equilateral triangle using the GSP

How can one construct an equilateral triangle? One can begin with the construction of a segment AB (Fig. 13.3, step 1). Designate its left endpoint (A) as the center of rotation (step 2) and rotate point (B) about point (A) by 60° (step 3). The rotation yields the third vertex (C) (step 4). One can see that this rotation of (B) about (A) along the arc BC creates a 60° angle CAB and because AC = AB, other two angles of triangle ABC measure 60° as well. Highlighting (in any order) the vertices (A), (B), (C), and selecting the line *Segment* in the *Construct* menu result in an equilateral triangle ABC (step 5).

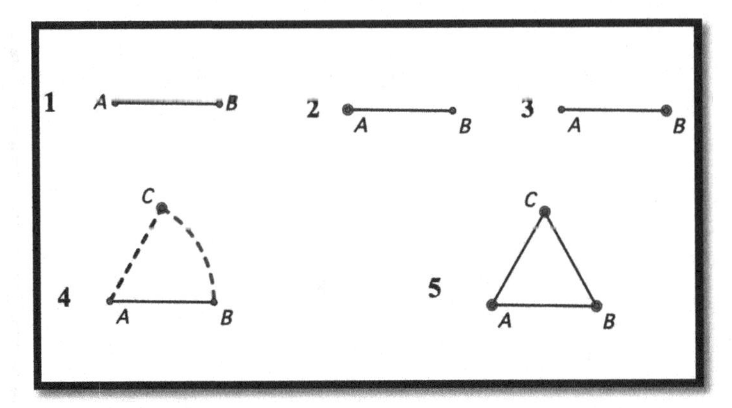

Fig. 13.3. Five steps leading to the construction of an equilateral triangle.

13.3 Constructing a square using the GSP

How can one construct a square? One can begin with the construction of a segment AB (Fig. 13.4, step 1). Designate its left endpoint (A) as the center of rotation (step 2) and click at point (B) (step 3) as the point to be rotated about point (A). The rotation of point (B) by 90° yields the third vertex (C) (step 4). Now, vertex (C) is used as a new center about which point (A) is rotated by 90° yielding point (D) (step 5), the fourth vertex. One can see that rotation of (B) about (A) by 90° along the arc BC creates a right angle CAB (step 4). Likewise, rotation of (A) about (C) by 90° (step 5) along the arc AD creates a right angle ACD. Highlighting the vertices (A), (C), (D), (B) – in that order, and selecting the line *Segment* in the *Construct* menu results in a square (step 6).

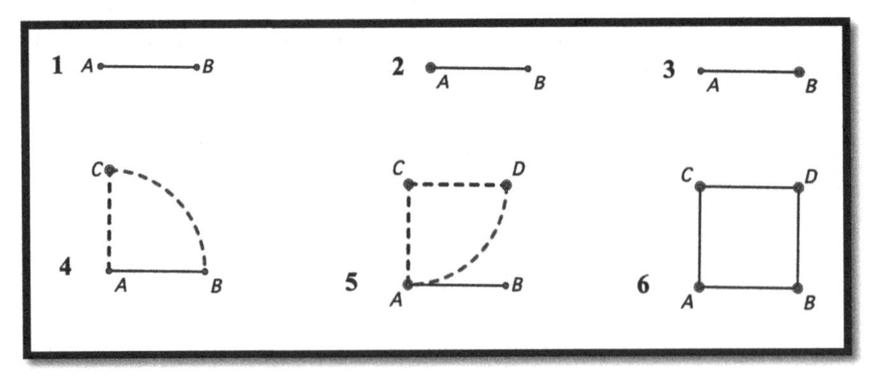

Fig. 13.4. Six steps leading to the construction of a square.

13.4 Constructing a regular pentagon using the GSP

How can one construct a regular pentagon? One can begin with the construction of a segment AB (Fig. 13.5, step 1). Designate its left endpoint (A) as the center of rotation (step 2) and highlight point B (step 3) as the point to be rotated about point A. The rotation of (B) by 108° $\left(=\frac{180^\circ(5-2)}{5}\right)$ about (A) yields vertex (C) (step 4). Now, vertex (C) is used as a new center about which point (A) has to be rotated by 108° yielding point (D) (step 5), the fourth vertex. Next, vertex (D) is used as a new center about which point (C) is rotated by 108° yielding point (E) (step 6), the fifth vertex. One can see that rotation of (B) about (A) by 108° along the arc BC creates an angle CAB (step 4) – one of the angles of a regular pentagon. Likewise, rotation of (A) about (C) by 108° along the arc AD and (C) about (D) by 108° along the arc CE create, respectively, angles ACD (step 5) and CDE (step 6), another two angles of a regular pentagon. Highlighting vertices (A), (C), (D), (E), (B) – in that order, and selecting the line *Segment* in the *Construct* menu results in a regular pentagon (step 7).

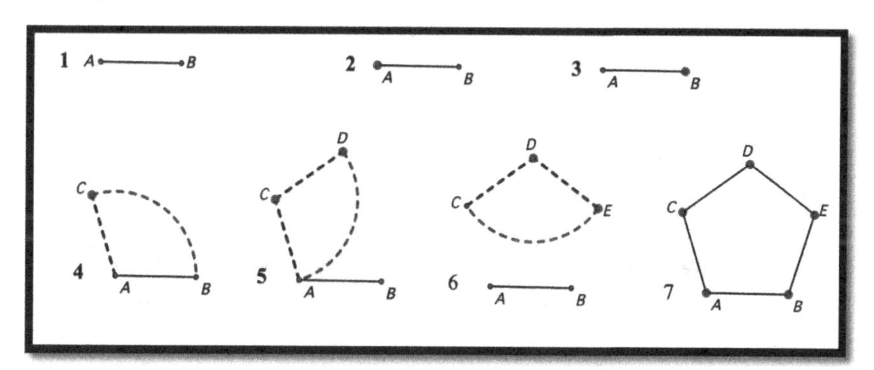

Fig. 13.5. Seven steps leading to the construction of a regular pentagon.

One can construct a regular pentagon using a different, more effective technique available in the context of GSP. One can begin with the construction of segment AB (Fig. 13.6). Designate its left endpoint (A) as the center of rotation and rotate point (B) about point (A) by 108°. The rotation yields vertex (C). Now, instead of assigning new roles to points (A) and (C), one can highlight points (B) and (A) – in that order, in the *Transform* menu choose *Iterate*, click at (A) and (C) – in that order, hit the button *Iterate* in the dialogue box of *Iterate*, and complete the pentagon construction by pressing the *plus* button on the keyboard. Note that the vertices developed through iteration are inactive; in particular, they cannot be labeled or used in further constructions. However, the method of constructing regular polygons through iteration proves to be very effective when the number of sides of a polygon is large. The angle of rotation in the construction of a regular polygon with n sides is equal to $\frac{180^\circ(n-2)}{n}$ where the numerator is the sum of all angles in a polygon with n sides (the formula was used above for $n = 5$).

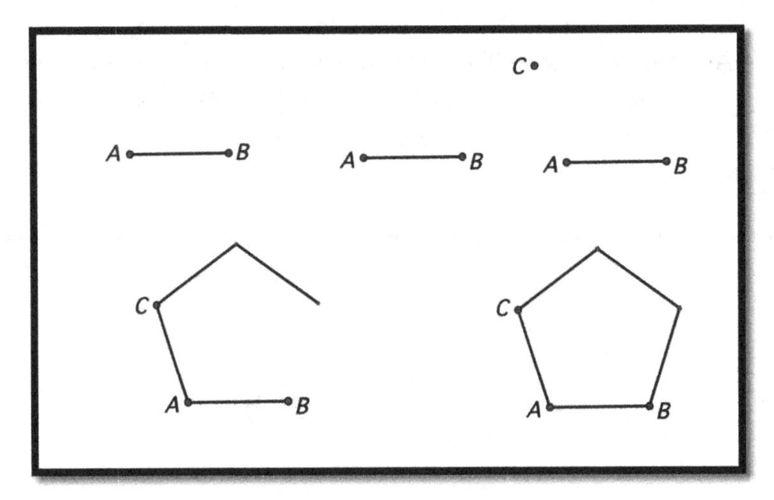

Fig. 13.6. Construction a regular pentagon through iteration.

13.5 Tessellation with triangles and quadrilaterals using the GSP

13.5.1 Tessellation with triangles

Aiming at the demonstration that tessellation can be carried out with scalene triangles, the following seven steps are presented with reference to Fig. 13.7.

Step 1. Construct a triangle.

Step 2. Use different line styles for each side of the triangle by highlighting a side, opening *Display* menu, opening *Line Style*, and selecting one of the three options: *Solid*, *Dashed*, *Dotted*.
Step 3. Highlight one of the sides (e.g., *Dashed*), open *Construct* menu and choose *Midpoint*.
Step 4. Double click at the constructed midpoint to designate it as the center of rotation of the triangle by 180°.
Step 5. Highlight the entire triangle, open *Transform* menu, select *Rotate*, enter 180 into the input box and click the button *Rotate*. This results in the appearance of a second triangle identical to the first one.
Step 6. Highlight the side adjacent to the one holding the previous center of rotation (as shown in Fig. 13.7), construct its midpoint and rotate the new triangle about that point by 180°. This results in the appearance of the third identical triangle.
Step 7. The three identical triangles form a trapezoid the larger base of which consists of two identical sides. These sides meet at one of the vertices of the original triangle. Rotate the three triangles about that vertex by 180°. As a result, the entire construction consists of six identical triangles covering space around a single point with no gaps or overlaps. One can pick up one of the vertices of the original triangle and drag it to see how all six triangles change, while keeping the effect of tessellation intact. This constitutes an informal demonstration (based on the dynamic feature of the software) that all triangles tessellate.

Remark 13.1. One can see that three angles formed by the pairs of sides (*Dotted*, *Dashed*), (*Dashed*, *Solid*) and (*Solid*, *Dotted*) add up to a straight angle. But these pairs of sides include three angles of the original triangle implying that the sum of three angles in a triangle is equal to 180° (see also Chapter 10, Fig. 10.1).

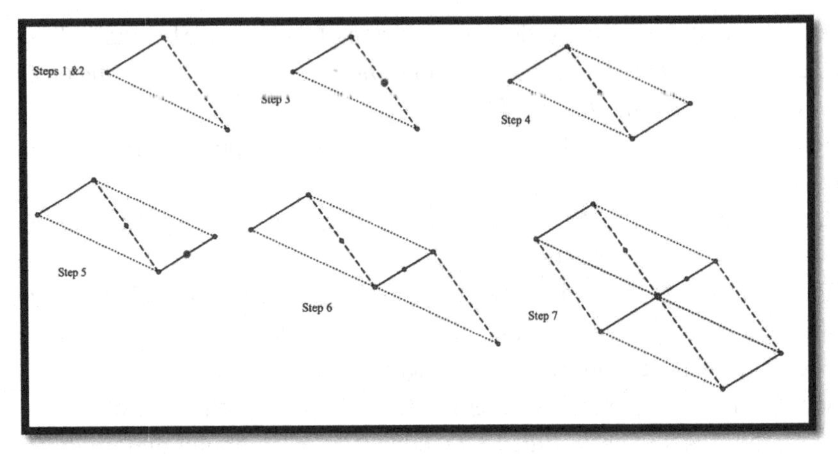

Fig. 13.7. Tessellation with triangles using the *Geometer's Sketchpad.*

13.5.2 Tessellation with quadrilaterals

Aiming at the demonstration that tessellation can be carried out with quadrilaterals, the following seven steps are presented with reference to Fig. 13.8.

Step 1. Construct a quadrilateral with sides having *Medium* line style.

Step 2. Use different line styles for each side of the quadrilateral by highlighting a side, opening *Display* menu, opening *Line Style*, and selecting one of the four options: *Thick, Solid, Dashed, Dotted.*

Step 3. Highlight one of the sides (e.g., *Solid*), open *Construct* menu and choose *Midpoint.*

Step 4. Double click at the constructed midpoint to designate it as the center of rotation of the quadrilateral by 180°.

Step 5. Highlight the entire quadrilateral, open *Transform* menu, select *Rotate*, enter 180 into the input box and click the button *Rotate.* This results in the appearance of a second quadrilateral identical to the first one.

Step 6. Highlight the side adjacent to the one holding the previous center of rotation (as shown in Fig. 13.8), construct its midpoint and rotate the second quadrilateral about that point by 180°. This results in the appearance of the third quadrilateral identical to the first two.

Step 7. Repeat the previous step by rotating the third quadrilateral by 180° about the midpoint of the side sharing a point with another two sides which hold the previous centers of rotation. As a result, the entire construction consists of four identical quadrilaterals covering space around a single point with no gaps or overlaps. One can pick up one of the vertices of the original quadrilateral and drag it to see how all four quadrilaterals change yet keeping the effect of tessellation intact. This constitutes the informal demonstration (based on the dynamic feature of the software) that all quadrilaterals tessellate.

Remark 13.2. One can see that four angles formed by the pairs of sides (*Solid, Thick*), (*Thick, Dotted*), (*Dotted, Dashed*) and (*Dashed, Solid*) add up to a 360° angle. But these pairs of sides include four angles of the original quadrilateral implying that the sum of four internal angles of a quadrilateral is equal to 360°.

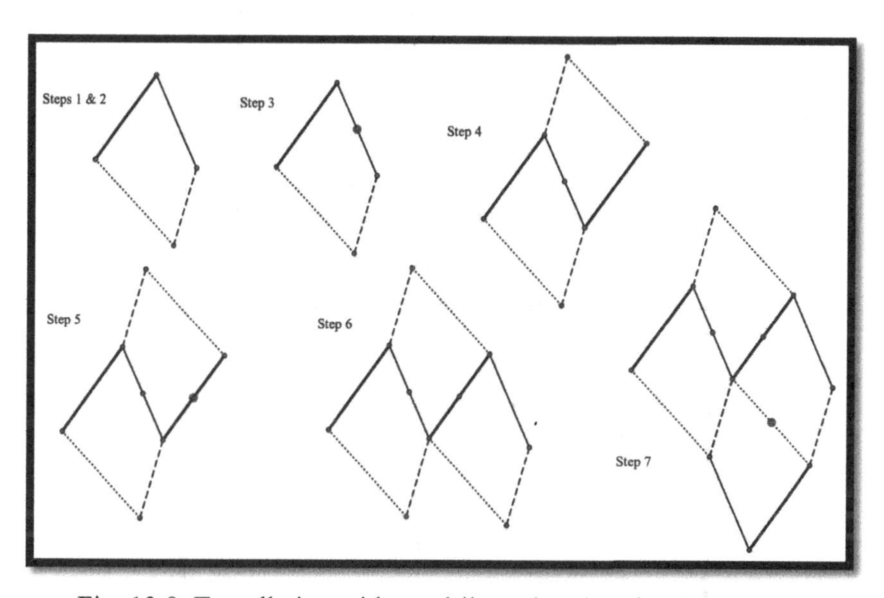

Fig. 13.8. Tessellation with quadrilaterals using the *Geometer's Sketchpad.*

13.6 Programming of the *Cookies on Plates* spreadsheet

The following programming details are presented with reference to Fig. 13.9.

In column A (beginning from cell A3) consecutive counting numbers are defined.

Cell B3: =IF(AND(INT((B$1-A3)/2)=(B$1-A3)/2,B$1-A3>0),(B$1-A3)/2," ").

Attaching a scroll bar to cell B1 makes it possible to have a dynamic spreadsheet which displays all pairs of cookies located on the first two plates to have, through the ATLT rule, on the fourth plate the number of cookies displayed in cell B1.

	A	B	C
1	◄ ►	10	4th plate
2	1st plate	2nd plate	
3	1		
4	2	4	
5	3		
6	4	3	
7	5		
8	6	2	
9	7		
10	8	1	
11	9		

Fig. 13.9. Spreadsheet Cookies on Plates.

	A	B	C	D	E
1	cookies	25	plates	5	
2		◄ ►		◄ ►	
3	1st plate	2nd plate			
4	1				
5	2	7	9	16	25
6	3				
7	4				
8	5	5	10	15	25
9	6				
10	7				
11	8	3	11	14	25
12	9				
13	10				
14	11	1	12	13	25

Fig. 13.10. Extended spreadsheet Cookies on Plates.

The following programming details are presented with reference to Fig. 13.10.

In column A (beginning from cell A4) consecutive counting numbers are defined.

Cell F1: =D1 – 1, cell G1: =F1 – 1.
Cell H3: = (1/SQRT(5))*(((1+SQRT(5))/2)^F1-((1-SQRT(5))/2)^F1)
Cell I3: = (1/SQRT(5))*(((1+SQRT(5))/2)^G1-((1-SQRT(5))/2)^G1)
The last two formulas are based on the closed formula for the n-th Fibonacci number

$$F_n = \frac{1}{\sqrt{5}}\left[\left(\frac{1+\sqrt{5}}{2}\right)^n - \left(\frac{1-\sqrt{5}}{2}\right)^n\right], \qquad n = 1, 2, 3, \ldots.$$

Cell B4: =IF(AND(INT((B$1-I$3*A4)/H$3)=(B$1-I$3*A4)/H$3, B$1-I$3*A4>0),(B$1-I$3*A4)/H$3," ") – replicated down column B.
Cell C4: =IF(AND(COUNT(B4)>0, COUNT($A4:B4)<$D$1), A4+B4," ") – replicated down rows and across columns of the spreadsheet.

13.7 Programming of the *Postage* spreadsheet

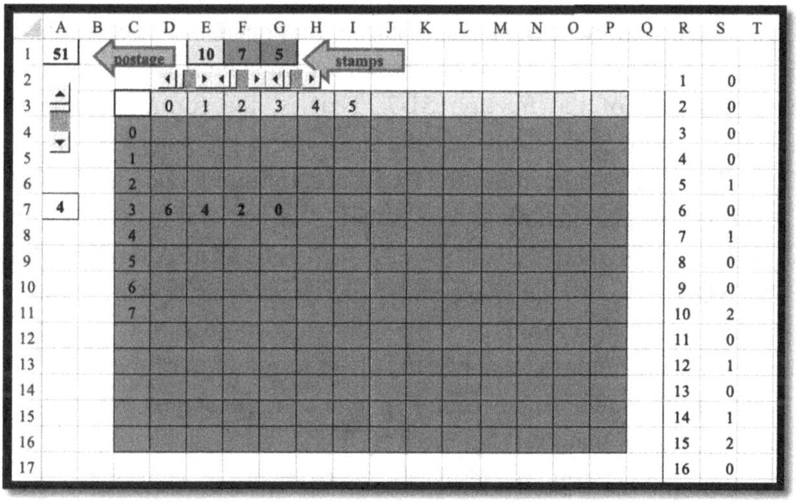

Fig. 13.11. Spreadsheet *Postage*.

The following programming details are presented with reference to Fig. 13.11. Cell A1 is slider-controlled and it displays postage required to send a letter. Cells E1, F1, and G1 are slider-controlled with numeric entries (denominations of stamps) displayed in the decreasing order. Cell D3: = 0; cell E3: =IF(D3<INT($A$1/$E$1), D3+1," ") – replicated to the right along row 3. The last formula enables the spreadsheet to terminate displaying the quantities of a stamp (the denomination of which is

displayed in cell E1) that would make the postage larger than needed. For example, for the 51-cent postage, the largest number of 10-cent stamps is determined by the integer part of the fraction 51/10. That is, the formula terminates displaying numbers in a cell of row 3 (and, instead, enters the symbol " " in this cell) if the previous cell displays INT(51/10) = 5. Note that in the spreadsheet of Fig. 13.10, cell J3 is entered with the symbol " " and all subsequent cells in row 3 do not display anything. This means that Excel programming code when verifying in cell H3 the inequality J3<INT($A$1/$E$1) with J3 marked with the symbol " ", does not recognize this inequality as being satisfied and, thereby, acts in accord with what the conditional function IF does when the condition is not satisfied; that is, marks the cell H3 with the symbol " ".

Cell C4: = 0; cell C5: =IF(C4<INT($A$1/$F$1), C4+1," ") – replicated down column C. This formula enables the spreadsheet to terminate displaying the quantities of a stamp the postage of which is displayed in cell F1 that would make the postage larger than needed. For example, for the 51-cent postage, the largest number of 7-cent stamps is determined by the integer part of the fraction 51/7. That is, the formula terminates displaying numbers in a cell of column C if the previous cell displays INT(51/7) = 7. The note about the Excel programming code made above for a similar formula including the symbol " " is applicable here as well.

Cell D4: =IF(OR(D$3=" ",$C4=" ")," ", IF(AND($A$1-D$3*$E$1-$C4*$F$1>=0, MOD($A$1-D$3*$E$1-$C4*$F$1,$G$1)=0),($A$1-D$3*$E$1-$C4*$F$1)/$G$1, " ")) – replicated across rows and down columns. This formula is designed to solve the equation of the form $ax + by + cz = d$ (where the values of a, b, c, and d are displayed, respectively, in cells E1, F1, G1, and A1) by

(i) isolating the term $cz = d - (ax + by)$,

(ii) selecting different pairs of x and y (displayed, respectively, in row 3 – beginning from cell D3, and column C – beginning from cell C4), and

(iii) verifying whether the expression $\frac{d-(ax+by)}{c}$ yields an integer value of z (to be displayed in the range D4:I11); in other words, verifying whether dividing $d - (ax + by)$ by c yields a zero remainder.

Cell A7: =COUNT(D4:P16) – it displays the number of ways the postage can be made with the stamps available.
Cell R2: =1; cell R3: = R2 + 1 – replicated down column R.
Cell S2: =IF(A$1=0," ",IF(R2=A$1,A$7,S2)) – replicated down column S. This formula uses a circular reference, i.e., reference to the cell in which the formula is defined. Such reference enables preservation of the display of the results of computations dependent on an absolute reference to a cell the value of which changes at any new step as computing moves from cell to cell using a formula with a circular reference.

13.8 Programming of the spreadsheet of Fig. 5.10

SIZE | DIVISIBLE BY | 2

10

	1	2	3	4	5	6	7	8	9	10
1	1		3		5		7		9	
2										
3	3		9		15		21		27	
4										
5	5		15		25		35		45	
6										
7	7		21		35		49		63	
8										
9	9		27		45		63		81	
10										

Fig. 13.12. Odd products in the 10 × 10 multiplication table.

Fig. 13.12 shows the multiplication table of a variable size controlled by the scroll bar attached to cell A2 (which is given name n used in the formulas defined in row 3 and column A). Changing the content of cell I1 via the corresponding scroll bar enables the spreadsheet, through the conditional function IF presented below, to display only those products which are NOT multiples of the integer displayed in cell I1. With the number 2 in cell I1, the spreadsheet displays odd products only. More specifically, the following programming allows for the appearance of only odd products in the multiplication table.

Cell B3: =1, cell C3: =IF(B3<n,1+B3," ") – replicated to the right of row 3. This formula makes it possible to limit factors displayed in row 3 to those not greater than the size of the table. For example, when $n = 10$, the formula terminates displaying factors in a cell of row 3 if the previous cell displays 10. For more details about such a formula coding impact see section 13.6.

Cell A4: =1, cell A5: =IF(A4<n,1+A4," ") – replicated down column A. This formula makes it possible to limit factors displayed in column A to those not greater than the size of the table. For example, when $n = 10$, the formula terminates displaying factors in a cell of column A if the previous cell displays 10. For more details about such a formula coding impact see section 13.7.

Cell B4: =IF(OR(B$3=" ",$A4=" ")," ", IF(MOD(B$3*$A4,$I$1) =1,B$3*$A4," ")) – replicated across rows and down columns. This formula limits the multiplication table to the use of factors not greater than the given size of the table (the condition of the external function IF with reference to row 3 and column A) and for those factors displayed in row 3 and column A, through the use of internal function IF, displays their products only when they are not divisible by the number displayed in cell I1.

13.9 Programming of the spreadsheet of Fig. 5.11

The addition table shown in Fig. 13.13 has a variable size controlled by the scroll bar attached to cell A2 (which is given name n used in the formulas defined in row 3 and column A). Changing the content of cell I1 via the corresponding scroll bar enables the spreadsheet to display only the multiples of the integer displayed in cell I1.

Cell B3: =1, cell C3: =IF(B3<n,1+B3," ") – replicated across row 3. Cell A4: = 1, cell A5: =IF(A4<n,1+A4," ") – replicated down column A. For specific details about the last two conditional formulas' coding impact see section 13.6.

Cell B4: =IF(OR(B$3=" ",$A4=" ")," ", IF(MOD(B$3+$A4,$I$1) =0,B$3+$A4, " ")) – replicated across columns and down rows. This formula limits the addition table to the use of addends not greater than the given size of the table (the condition of the external function IF with reference to row 3 and column A) and for those addends displayed in row

3 and column A, through the use of internal function IF, the spreadsheet displays their sums only when they are divisible by the number displayed in cell I1.

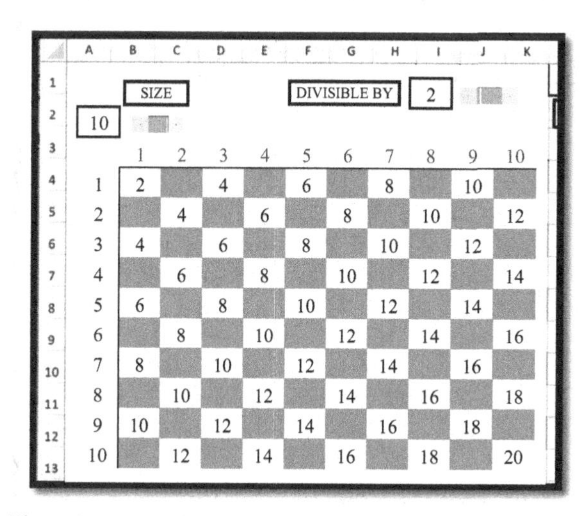

	1	2	3	4	5	6	7	8	9	10
1	2		4		6		8		10	
2		4		6		8		10		12
3	4		6		8		10		12	
4		6		8		10		12		14
5	6		8		10		12		14	
6		8		10		12		14		16
7	8		10		12		14		16	
8		10		12		14		16		18
9	10		12		14		16		18	
10		12		14		16		18		20

Fig. 13.13. Even sums in the 10 × 10 addition table.

13.10 Programming of the spreadsheet of Fig. 5.12

The addition table shown Fig. 13.14 has a variable size controlled by the scroll bar attached to cell A2 (which is given name n used in the formulas defined in row 3 and column A). Changing the content of cell I1 via the corresponding scroll bar enables the spreadsheet to display only the multiples of the integer entered in cell I1. Cell B3: =1, cell C3: =IF(B3<n,1+B3," ") – replicated across row 3. Cell A4: = 1, cell A5: =IF(A4<n,1+A4," ") – replicated down column A. For specific details about the last two conditional formulas' coding impact see section 13.6.

Cell B4: =IF(OR(B$3=" ",$A4=" ")," ", IF(MOD(B$3+$A4,$I$1) =1,B$3+$A4, " ")) – replicated across columns. This formula limits the addition table to the use of addends not greater than the given size of the table (the condition of the external function IF with reference to row 3 and column A) and for those addends displayed in row 3 and column A, through the use of internal function IF, displays their sums only when they are not divisible by the number displayed in cell I1.

A B C D E F G H I J K

SIZE DIVISIBLE BY 2

10

	1	2	3	4	5	6	7	8	9	10
1		3		5		7		9		11
2	3		5		7		9		11	
3		5		7		9		11		13
4	5		7		9		11		13	
5		7		9		11		13		15
6	7		9		11		13		15	
7		9		11		13		15		17
8	9		11		13		15		17	
9		11		13		15		17		19
10	11		13		15		17		19	

Fig. 13.14. Odd sums in the 10 × 10 addition table.

13.11 Programming of the spreadsheet of Fig. 10.19

The following programming details are presented with reference to Fig. 13.15. Cell A3 is controlled by a slider; cell B3: =4*A3; cell B4: =3 (the smallest side length allowing for rectangle with a hole); cell B5: =IF(OR(B4=" ", B4=C4), " ", B4+1) – replicated down column B (this formula increases the smaller side length by one until adjacent sides become congruent, that is, square is reached); cell C4: =IF(B4=" ", " ", 0.5*(B$3-2*B4)+2) – replicated down column C (this formula calculates a larger side length until square is reached); cell C3: =COUNT(B4:B36) – this formula counts the number of rectangles with a hole; cell D4: =IF(B4=" ", " ", B$3) – replicated down column D (this formula displays area of a rectangle); cell E4: =IF(D4=" "," ",4*(C4+B4-2)) – replicated down column E (this formula calculates perimeter of a rectangle).

	A	B	C	D	E
1					
2	*n*	4*n*	Solutions	Area	Perimeter
3	5	20	4		
4		3	9	20	40
5		4	8	20	40
6		5	7	20	40
7		6	6	20	40

Fig. 13.15. There are four rectangles with a hole having area 20 and perimeter 40.

13.12 Programming of the *Tossing Four Coins* spreadsheet

The following programming details are presented with reference to Fig. 13.16. Cells A3, B3, C3, D3: =RANDBETWEEN(0,1) – replicated down to row 10002. A two-argument function RANDBETWEEN(a, b) generates a random integer in the range [*a*, *b*] including *a* and *b*. In particular, for the range [0, 1] the function randomly generates either 0 or 1. Cell F3: =IF(SUM(A3:D3)=4,"HHHH", IF(AND(SUM(A3:D3)=2, A3=1,C3=1),"HTHT"," ")) – replicated to cell F10002. Assuming that 0 and 1 stand, respectively, for T and H, the formula determines when each of the four tosses yields H and if so, it displays the symbol HHHH, otherwise (looking for the case HTHT, decoded as 1010) the internal conditional function IF adds the numbers displayed in the range [A3:D3] and if the sum is 2, provided that cells A3 and C3 include 1's, it displays the symbol HTHT; otherwise, it leaves cell blank.

The next action is to count the number of HHHH's and HTHT's. To this end, a two-argument COUNTIF (range, what to count) function which displays the number of times the second argument (what to count) appears in the range defined by the first argument (range) can be used. That is, cell G1: =COUNTIF(F3:F10002,"HHHH"), cell H1: =COUNTIF(F3:F10002,"HTHT"). Finally, experimental probabilities of each of the two outcomes and their absolute difference are calculated, respectively, as follows: cell G2: =G1/10000, cell H2: =H1/10000, cell F2: =ABS(G2 – H2).

	A	B	C	D	E	F	G	H
1							596	595
2						0.0001	0.0596	0.0595
3	1	1	1	1		HHHH		
4	1	0	0	1				
5	0	1	1	1				
6	0	0	0	1				
7	0	1	0	1				
8	0	1	1	0				
9	0	1	1	1				
10	1	1	0	0				
11	1	1	0	0				
12	0	0	1	1				
9998	1	1	1	1		HHHH		
9999	0	1	1	0				
10000	1	0	1	0		HTHT		
10001	0	1	0	1				
10002	1	0	0	0				

Fig. 13.16. Tossing Four Coins.

13.13 Programming of the spreadsheet of Fig. 12.10

The following programming details are presented with reference to Fig. 13.17. Cells A2, B2, C3: =RANDBETWEEN(1,6) – replicated down to row 5001. Cell D2: =IF(SUM(A2:C2)=10, 1, 0) – replicated to cell D5001. Cell E1: = SUM(D2:D5001)/5000.

	A	B	C	D	E
1	Die 1	Die 2	Die 3		0.1246
2	5	4	4	0	
3	4	2	6	0	
4	1	3	3	0	
5	2	5	5	0	
6	3	5	4	0	
7	6	2	3	0	
4999	1	6	4	0	
5000	3	2	5	1	
5001	1	5	2	0	

Fig. 13.17. Experimental probability (cell E1) to cast 10 when rolling three dice 5000 times.

13.14 Programming of the M&M spreadsheet of Figures 12.7 and 12.8
The following programming details are presented with reference to Fig. 13.18. Cells in the range [A5:D9] are slider-controlled and they display the quantities of plain [A5:A9] and peanut [D5:D9] M&Ms in five colors. Cell A1: = SUM(B5:B9) – it displays the total number of plain M&Ms, cell D1: = SUM(C5:C9) – it displays the total number of peanut M&Ms. Cell F5: =B9/A1 – it is hidden and calculates the ratio of red plain M&Ms to the total number of plain M&Ms. Cell H5: = C9/D1 – it is hidden and calculates the ratio of red peanut M&Ms to the total number of peanut M&Ms. The two charts, Plain and Peanut, are representing the values calculated in cells F5 and H5, respectively. To construct a chart representing numeric information in a cell, click at the cell, go to *Insert* menu, open *Chart* and select *Column*.

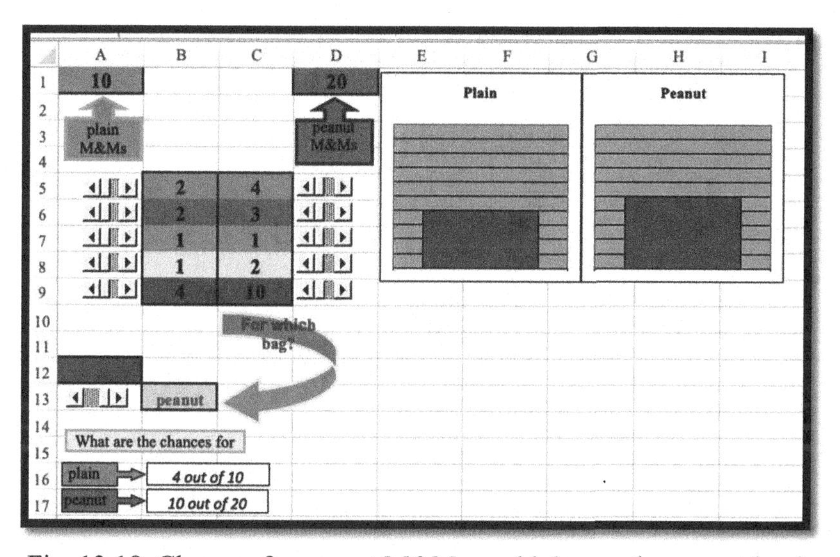

Fig. 13.18. Chances for peanut M&Ms are higher as the peanut bar is higher.

APPENDIX:
ACTIVITY SETS: 300 PROBLEMS AND QUESTIONS

ACTIVITY SET 1

Conceptual shortcuts, the order of operations, and arithmetic on place value charts

1) How can one find the value of the numeric expression $53 + 9 - 7$ through a conceptual shortcut? Explain your answer.
2) How can one find the value of the numeric expression $37 + 9 - 6$ through a conceptual shortcut? Explain your answer.
3) How can one find the value of the numeric expression $63 + 18 - 16$ through a conceptual shortcut? Explain your answer.
4) How can one find the value of the numeric expression $185 + 69 - 65$ through a conceptual shortcut? Explain your answer.
5) Formulate four arithmetical problems similar to 1) – 4) to be solved through a conceptual shortcut. Solve your problems.
6) Without recourse to the multiplication table, carry out the multiplication 5×7 (by defining multiplication as repeated addition, the first factor shows the number of repetitions of the second factor) through a conceptual shortcut using commutative property of multiplication. Explain the meaning of the conceptual shortcut used here.
7) Without recourse to the multiplication table, carry out the multiplication 10×7 (by defining multiplication as repeated addition, the first factor shows the number of repetitions of the second factor) through a conceptual shortcut using commutative property of multiplication. Explain the meaning of the conceptual shortcut used here.
8) Multiply 55 by 11 through a conceptual shortcut using a geometric representation of the product. Draw a picture.
9) Multiply 43 by 21 through a conceptual shortcut using a geometric representation of the product. Draw a picture.
10) Find the sum $1 + 2 + 3 + ... + 9 + 10$ through a conceptual shortcut. Demonstrate (on a picture, perhaps using a spreadsheet) how to use manipulatives in support of the conceptual shortcut. Connect this summation problem to the history of mathematics.
11) Find the sum $1 + 3 + 5 + ... + 17 + 19$ through a conceptual shortcut. Demonstrate (on a picture, perhaps using a spreadsheet) how to use manipulatives in support of the conceptual shortcut. Try to rearrange manipulatives on your picture to have a square. Determine the size of the square and connect it to the number of addends in the sum.

12) Find the sum $1+5+9+13$ by reducing it to the sum of four equal integers. Draw a picture (perhaps using a spreadsheet) describing your action.

13) Represent the sum $1+5+9+13+17$ as a product of two integers greater than one (do not consider different order of factors). Can this be done in more than one way? Why or why not?

14) Represent the sum $13+14+15+16$ as a product of two integers greater than one (do not consider different order of factors). Can this be done in more than one way? Why or why not?

15) Represent the sum $3+8+13+18+23$ as a product of two integers greater than one (do not consider different order of factors). Can this be done in more than one way? Why or why not?

16) Is it possible to represent the sum $4+9+14+19+25$ as a product of two integers greater than one? Compare this sum to those in the previous three examples. What difference do you see?

17) Pose and solve a problem when a sum of five integers (with a common difference) can be represented as a product of two factors greater than one in one way only (do not consider different order of factors).

18) Pose and solve a problem when a sum of five integers (with a common difference) can be represented as a product of two factors greater than one in more than one way (do not consider different order of factors).

19) Pose and solve a problem when a sum of five integers (with a common difference) can be represented as a product of two factors greater than one in only one way (do not consider different order of factors).

20) Pose a problem when a sum of five integers cannot be represented as a product of two factors greater than one. Explain your reasoning.

21) Pose and solve a problem when a sum of six integers cannot be represented as a product of two factors greater than one.

22) Find the sum $1+4+7+\ldots+55+58$ through a conceptual shortcut.

23) Two drimps and seven grimps have the total of twenty legs. Using either a conceptual shortcut or picture-supported trial and error, find the number of legs each creature, drimp and grimp, has. Show and explain your answer.

24) Three drimps and five grimps have the total of twenty-one legs. Using either a conceptual shortcut or picture-supported trial and error, find the number of legs each creature, drimp and grimp, has. Show and explain your answer.

25) Using pictures, solve the following problem. If 2 twiggles and 5 glogs have 18 legs, how many legs does each creature have? Alternatively, if (in a tax-free store) one spent

\$18 on buying notebooks and pens each of which costs \$2 and \$5, respectively, how many notebooks and how many pens did one buy? Now, solve the equation $2x + 5y = 18$ in whole numbers using a conceptual shortcut. Here the variables x and y stand for the number of legs each creature has; alternatively, x and y are the prices of a notebook and a pen. What if we replace 18 by 24? Does the corresponding equation (with 24 in the right-hand side) have only one solution? Why or why not?

26) Using pictures, solve the following problem. If 3 twiggles and 4 glogs have 20 legs, how many legs does each creature have? Alternatively, if (in a tax-free store) one spent \$20 on buying notebooks and pens each of which costs \$3 and \$4, respectively, how many notebooks and how many pens did one buy? Now, solve the equation $3x + 4y = 20$ in whole numbers using a conceptual shortcut. Here the variables x and y stand for the number of legs each creature has; alternatively, x and y are the prices of a notebook and a pen. What if we replace 20 by 28? Does the corresponding equation (with 28 in the right-hand side) have only one solution? Why or why not?

27) Using pictures, solve the following problem. If 3 twiggles and 2 glogs have 21 legs, how many legs does each creature have? Does the problem have more than one correct answer? Alternatively, if (in a tax-free store) one spent \$21 on buying notebooks and pens each of which costs \$3 and \$2, respectively, how many notebooks and how many pens did one buy? Now, solve the equation $3x + 2y = 21$ in whole numbers using a conceptual shortcut. Here the variables x and y stand for the number of legs each creature has; alternatively, x and y are the prices of a notebook and a pen. What if we replace 21 by 10? Does the corresponding equation (with 10 in the right-hand side) have more than one solution? Why or why not?

28) Solve the following problem through a conceptual shortcut. If, preparing for a party, one spent \$150 on water sets and pizzas each of which costs \$10 and \$13, respectively, how many water sets and how many pizzas did one buy? Consider a tax-free situation.

29) Solve the following problem through a conceptual shortcut. If, preparing for a party, one spent \$135 on water sets and pizzas each of which costs \$8 and \$9, respectively, how many sets of water and how many pizzas did one buy? Consider a tax-free situation.

30) Formulate a similar – to 25) and 26) – word problem that leads to an equation that can be solved through a conceptual shortcut.

31) Find the values of the following numeric expressions using a calculator, a spreadsheet and *Wolfram Alpha* (make sure each tool yields the same answer despite using different codes): $32 \div 8 + 24$; $32 \div (8 + 24)$; $24 \cdot 2 \div 2$; $24 \cdot (2 \div 2)$; $64 \cdot 16 \div 4$; $64 \div 16 \cdot (10 + 6)$; $64 \div 16 \cdot (6 - 2)$.

32) Find the sum $23 + 16$ using a place value chart.

33) Find the sum $27 + 35$ using a place value chart.

34) Find the difference $26 - 13$ using a place value chart.

35) Find the difference $35 - 17$ using a place value chart.

ACTIVITY SET 2

Connecting mathematical practice with mathematical content of additive decompositions of integers

1) If the key 3 on your calculator is broken, how could you find the sum 354 + 531 + 453? Is there more than one way of doing that? Show how this problem can be presented on a place value chart using multicolored counters (or square tiles).
2) If the keys 1 and 3 on your calculator are broken, how could you find the sum 111 + 232 + 311? Is there more than one way of doing that? Show how this problem can be presented on a place value chart using multicolored counters (or square tiles).
3) Assuming that you do need a calculator for adding two two-digit numbers and the keys 3, 4 and 8 do not work, how could you find the sum 13 + 31 using two positive integer addends only? How many different representations of this sum through two (digits 3, 4, and 8 free) positive integers are there? Write down all such representations both through the base-ten blocks and numerically. Do the same for the sum 12 + 21 with the key 2 broken. Do not consider different orders in which two numbers make up the sums.
4) Assuming that you do need a calculator for adding two two-digit numbers and the keys 2, 3 and 5 were broken, how could you find the sum 23 + 31 using two positive integer addends only? How many different representations of this sum through two positive integers (digits 2, 3 and 5 free) are there? Write down all such representations both through the base-ten blocks and numerically. Do not consider different orders in which two numbers make up the sum.
5) If the key 2 on your calculator is broken, how would you find the sum 12 + 19 still using the calculator and two integer addends only? List all the sums of two integers that may be entered into your calculator to get the right answer. Do not consider different orders in which two numbers make up the sum.
6) If the keys 1, 2, 3 on your calculator do not work, how would you find the sum 61 + 16 still using the calculator and two integer addends only? List all the sums of two integers that may be entered into your calculator to get the right answer. Do not consider different orders in which two numbers make up the sum.
7) If the key 2 on your calculator is broken, how would you find the sum 11 + 21 still using the calculator and two integer addends only? List all the sums of two integers that may be entered into your calculator to get the right answer. Do not consider different orders in which two numbers make up the sum. Represent your solution both using pictures and numerically.

8) How many ways can one put eight (identical) cookies on two plates without regard to the order of the plates, no plate having three cookies, and with at least one cookie on each plate? Represent your solution both using pictures and numerically.
9) How many ways can one put eight (identical) cookies on three plates without regard to the order of the plates, no plate having two cookies, and with at least one cookie on each plate? Represent your solution both using pictures and numerically.
10) How many ways can one put ten (identical) candies on three plates without regard to the order of the plates, no plate having either two or three candies, and with at least one candy on each plate? Represent your solution both using pictures and numerically.
11) How many ways can one put ten (identical) cookies on four plates without regard to the order of the plates, no plate having three cookies, and with at least one cookie on each plate? Represent your solution both using pictures and numerically.
12) Find all ways to put eleven (identical) cookies on three plates without regard to the order of the plates if each plate must have more than one cookie. Represent your solution both using pictures and numerically.
13) How many ways can one put seven (identical) candies on three plates without regard to the order of the plates allowing no candy on a plate? Represent your solution both using pictures and numerically.
14) How can one put five (identical) cookies on three different color (red, blue, white) plates with the red plate always having one cookie and allowing no cookie on other plates? Represent your solution both using pictures and numerically.
15) How can one place six (identical) candies in three different size (large, middle, small) cups with the large cup always having two candies and with at least one candy in other cups? Represent your solution both using pictures and numerically.
16) How can one place six (identical) candies in three different size (large, middle, small) cups with only the large cup having two candies and with at least one candy in other cups? Represent your solution both using pictures and numerically.
17) Three creatures – brimp, drimp, and grimp – have ten legs among them. If neither is a one-legged creature and brimp has fewer legs than drimp while drimp has fewer legs than grimp, how many legs does each creature have? Use a picture to solve this problem. Does the problem have more than one correct answer?
18) How many ways can one change a dime into nickels and pennies? Draw a chart to present your solution.
19) How many ways can one change a quarter into dimes, nickels, and pennies? Draw a chart to present your solution.

20) How many ways can one change a half dollar into quarters, dimes, and nickels? Draw a chart to present your solution.

21) How many ways can one change a dollar into half dollars, quarters, and dimes? Draw a chart to present your solution.

22) How many ways can one change a dollar into half dollars, quarters, and nickels? Draw a chart to present your solution.

23) Find all ways to add consecutive positive integers in order to reach sums in the range one through sixteen. (For example, 3, 4, and 5 are consecutive integers and their sum is equal to 12. There are also sums comprised of two, four, and five addends; e.g., $1 + 2 + 3 + 4 + 5 = 15$).

   a. Use a spreadsheet to make a table displaying all the sums.
   b. Which integers (sums) appear in the table more than one time?
   c. Which integers from the range [1, 16] do not appear in this table and what is special about them?
   d. What can be said about the sums of two consecutive counting numbers?
   e. What can be said about the sums of three consecutive counting numbers?
   f. What can be said about the sums of four consecutive counting numbers?
   g. Use manipulatives (e.g., square tiles) to support your answers to questions d – f.
   h. Draw trapezoid-like representations for the numbers 9 and 15 through the sums of consecutive positive integers using counters.

24) Among the following four statements, a – d, only one statement is true. Which one is true? Explain your answer by analyzing each of the four statements.

   a. The sum of any three consecutive counting numbers is not divisible by three.
   b. The sum of any four consecutive counting numbers is divisible by four.
   c. Any number greater than one and smaller than 20 can be represented as a sum of consecutive counting numbers.
   d. There is only one counting number smaller than 10 which can be decomposed into a sum of consecutive counting numbers in more than one way.

30) Among the following four statements, a – d, only one statement is true. Which one is true? Explain your answer by analyzing each of the four statements.

   a. The sum of any two consecutive counting numbers is divisible by two.
   b. The sum of any six consecutive counting numbers is divisible by six.
   c. The sum of three consecutive counting numbers is not divisible by three.
   d. There is only one two-digit number smaller than 16 which can be decomposed into a sum of consecutive counting numbers in more than one way.

25) Among the following four statements, a – d, only one statement is true. Which one is true? Explain your answer by analyzing each of the four statements.

a. The sum of three consecutive counting numbers is always an even number.

b. The sum of three consecutive counting numbers is always an odd number.

c. The sum of any five consecutive counting numbers is not divisible by five.

d. There is only one two digit number smaller than 21 which can be decomposed into a sum of consecutive counting numbers in more than one way.

ACTIVITY SET 3

Explorations with multiplication and addition tables

1) Use *Wolfram Alpha* to construct a 5×5 multiplication table (type in the input box of the program "5 multiplication table").
2) Show five ways to find the sum of all numbers in the 5×5 multiplication table.
3) Using any of the five ways, find the sums of all numbers in the 4×4, 3×3, 2×2, and 1×1 multiplication tables.
4) Make an organized list of the (five) sums and enter it in the input box of *Wolfram Alpha* in order to find the continuation of this sequence. What is the first number generated by *Wolfram Alpha* in response?
5) Use one of the ways of finding the sum in item 2) in order to find the sum of numbers in the 6×6 multiplication table and check if your finding corresponds to the number generated by *Wolfram Alpha.*
6) Find the sum of numbers in the 10×10 multiplication table. How can one use *Wolfram Alpha* to check the answer?
7) Find the sum of numbers in the $n \times n$ multiplication table.
8) It is known that the sum of numbers in the largest gnomon of a multiplication table is equal to 216. Find the sum of all numbers in this table.
9) Use *Wolfram Alpha* to construct the 5×5 addition table (type in the input box of the program "5 addition table").
10) Show five ways to find the sum of all numbers in the 5×5 addition table.
11) Using any of the five ways, find the sums of all numbers in the 4×4, 3×3, 2×2, and 1×1 addition tables.
12) Make an organized list of the (five) sums and enter it in the input box of *Wolfram Alpha* in order to find the continuation of this sequence. What is the first number generated by *Wolfram Alpha* in response?
13) Use one of the ways of finding the sum in item 10) in order to find the sum of numbers in the 6×6 addition table and check if your finding corresponds to the number generated by *Wolfram Alpha.*
14) Find the sum of numbers in the 10×10 addition table. How can one use *Wolfram Alpha* to check the answer?
15) Find the sum of numbers in the $n \times n$ addition table.
16) How many even numbers are there in the 6×6 multiplication table?
17) How many odd numbers are there in the 6×6 multiplication table?

18) How many multiples of three are there in the 6 × 6 multiplication table?
19) How many even numbers are there in the 6 × 6 addition table?
20) How many odd numbers are there in the 6 × 6 addition table?
21) How many multiples of three are there in the 6 × 6 addition table?
22) Why is the sum of numbers located on the main (top-left/bottom-right) diagonal of the 6 × 6 addition table equal twice the sum of numbers located in the first row of the 6 × 6 multiplication table?
23) How can one develop in the 6 × 6 multiplication table the sequence 1, 4, 10, 20, 35, 56?
24) How can one prove without direct computational verification that $1 + 4 + 10 + 20 + 35 + 56 + 70 + 76 + 73 + 60 + 36 = (1 + 2 + 3 + 4 + 5 + 6)^2$?
25) Enter the numbers 1, 4, 10, 20, 35, 56 into the input box of *Wolfram Alpha.* The program shows that the next number in this sequence is 84. What is the meaning of 84 as the seventh number in this sequence in the context of multiplication tables?
26) *Wolfram Alpha* provides the fraction $\frac{n(n+1)(n+2)}{6}$ as the general term of the sequence 1, 4, 10, 20, 35, 56, 84, ... How can one explain that this fraction is always an integer?
27) Consider the 6 × 6 addition table. How can one develop the sequence 2, 6, 12, 20, 30, 42 in this table?
28) Enter the numbers 2, 6, 12, 20, 30, 42 into the input box of *Wolfram Alpha.* The program shows that the next number in this sequence is 56. What is the meaning of the number 56 as the seventh number in this sequence in the context of addition tables?
29) Each term of the sequence 2, 6, 12, 20, 30, 42 developed within the 6 × 6 addition table is twice the corresponding term of the sequence of triangular numbers 1, 3, 6, 10, 15, 21 (which are partial sums of consecutive counting numbers 1, 2, 3, 4, 5, 6). How can this connection between the addition table and the triangular numbers be explained?
30) The sequence 2, 6, 12, 20, 30 can also be found in the 6 × 6 multiplication table as entries of the diagonal which is immediately above the main (top-left/bottom-right) diagonal of the table. How can this connection between the addition and the multiplication tables be explained?
31) Find the sum of even numbers in the 6 × 6 multiplication table.
32) Find the sum of odd numbers in the 6 × 6 multiplication table.
33) Find the sum of multiples of three in the 6 × 6 multiplication table.
34) Find the sum of even numbers in the 6 × 6 addition table.
35) Find the sum of odd numbers in the 6 × 6 addition table.
36) Find the sum of multiples of three in the 6 × 6 addition table.

37) Find the sum of even numbers in the $n \times n$ multiplication table.

38) Find the sum of odd numbers in the $n \times n$ multiplication table.

39) Find the sum of even numbers in the $n \times n$ addition table.

40) Find the sum of odd numbers in the $n \times n$ addition table.

ACTIVITY SET 4

Using technology in solving and posing problems

1) Following the ATLT rule (Chapter 6, Section 6.4), 15 cookies were put on the 5th plate. Using the spreadsheet Cookies on Plates (see Chapter 13, Section 13.6), find all possible numbers of cookies on the first two plates. How can one repeat the first and the second plates with cookies to have the total of 15 cookies? Draw a diagram with the total of 15 cookies on the two types of plates. Construct an equation to solve the problem using a conceptual shortcut (Chapter 3, Section 3.4).
2) Following the ATLT rule (Chapter 6, Section 6.4), 21 cookies were put on the 5th plate. Using the spreadsheet Cookies on Plates (see Chapter 13, Section 13.6), find all possible numbers of cookies on the first two plates so repeated. How can one repeat the first and the second plates with cookies to have the total of 21 cookies? Draw a diagram with the total of 21 cookies on the two types of plates. Construct an equation to solve the problem using a conceptual shortcut (Chapter 3, Section 3.4).
3) Following the ATLT rule (Chapter 6, Section 6.4), 24 cookies were put on the 6th plate. Using the spreadsheet Cookies on Plates (see Chapter 13, Section 13.6), find all possible numbers of cookies on the first two plates so repeated. How can one repeat the first and the second plates with cookies to have the total of 24 cookies? Draw a diagram with the total of 24 cookies on the two types of plates. Construct an equation to solve the problem using a conceptual shortcut (Chapter 3, Section 3.4).
4) Following the ATLT rule (Chapter 6, Section 6.4), 90 cookies were put on the 7th plate. Using the spreadsheet Cookies on Plates (see Chapter 13, Section 13.6), find all possible numbers of cookies on the first two plates. How can one repeat the first and the second plates with cookies to have the total of 90 cookies on the two types of plates? Construct an equation to solve the problem using a conceptual shortcut (Chapter 3, Section 3.4).
5) Following the ATLT rule (Chapter 6, Section 6.4), 65 cookies were put on the 6th plate. Using the spreadsheet Cookies on Plates (see Chapter 13, Section 13.6), find all possible numbers of cookies on the first two plates. How can one repeat the first and the second plates with cookies to have the total of 65 cookies? Construct an equation to solve the problem using a conceptual shortcut (Chapter 3, Section 3.4).
6) Following the ATLT rule (Chapter 6, Section 6.4), 90 cookies were put on the 8th plate. Is there more than one way to do that? Why or why not? How can one repeat the first and the second plates with cookies to have the total of 90 cookies? Draw a diagram with the total

of 90 cookies on the two types of plates. Construct an equation to solve the problem using a conceptual shortcut (Chapter 3, Section 3.4).

7) Is it possible by using the ATLT rule (Chapter 6, Section 6.4), to put 30 cookies on the 7th plate starting with the first two plates? Why or why not?
8) Is it possible by using the ATLT rule (Chapter 6, Section 6.4), to put 40 cookies on the 8th plate starting with the first two plates. Why or why not?
9) Is it possible by using the ATLT rule (Chapter 6, Section 6.4), to put 63 cookies on the 9th plate starting with the first two plates. Why or why not?
10) Using the spreadsheet Cookies on Plates (see Chapter 13, Section 13.6), formulate and solve problems similar to problems 1 – 6.
11) Using the spreadsheet Cookies on Plates (see Chapter 13, Section 13.6), formulate and solve problems similar to problems 7 – 9.
12) It takes 46 cents in postage to mail a letter. A post office has stamps of denomination 5 cents, 7 cents, and 10 cents. How many combinations of the stamps could Ada buy to send a letter if the order in which the stamps are arranged on an envelope does not matter? What are those combinations? How can you prove that you found all combinations of the stamps?
13) It takes 41 cents in postage to mail a letter. A post office has stamps of denomination 5 cents, 7 cents, and 10 cents. How many combinations of the stamps could Alan buy to send a letter if the order in which the stamps are arranged on an envelope does not matter? What are those combinations? How can you prove that you found all combinations of the stamps?
14) Use the spreadsheet Postage (see Chapter 13, Section 13.7) to generate and formulate a problem similar to Problems 12 and 13. List all solutions to your own problem. Develop and explain a systematic way of solving your problem when technology (the spreadsheet) is not available. Note: a teacher uses technology to pose a problem for students to solve it without technology. Answer the following questions related to the problem you posed.
    a. How many solutions (different answers) does your problem have?
    b. Do you think that the spreadsheet generated all solutions to your problem? Why or why not?
    c. What grade level(s) is your problem appropriate for?
    d. Do you expect young children to find all solutions to your problem? Explain your answer.
    e. Is your problem numerically coherent (Chapter 6, Section 6.2)? Why or why not?
    f. Is your problem pedagogically coherent (Chapter 6, Section 6.2)? Why or why not?

g. Is your problem contextually coherent (Chapter 6, Section 6.2)? Why of why not?

h. What do you think about the role that computing technology can play in problem posing by elementary teachers?

ACTIVITY SET 5

Learning to move from visual to symbolic

1) Using the diagram of Fig. AS 1 show that $5 \times 1 + 5 \times 10 = 5 \times 2 + 5 \times 9 = 5 \times 3 + 5 \times 8$. What is special about the pairs of numbers (1, 10), (2, 9) and (3, 8)?
2) Using the diagram of Fig. AS 2 show that $(1 + 7) + (10 + 7) = (2 + 7) + (9 + 7) = (3 + 7) (8 + 7)$. What do the pairs of numbers (8, 17), (9, 16) and (10, 15) have in common?

×	1	2	3	4	5	6	7	8	9	10
1	1	2	3	4	5	6	7	8	9	10
2	2	4	6	8	10	12	14	16	18	20
3	3	6	9	12	15	18	21	24	27	30
4	4	8	12	16	20	24	28	32	36	40
5	5	10	15	20	25	30	35	40	45	50
6	6	12	18	24	30	36	42	48	54	60
7	7	14	21	28	35	42	49	56	63	70
8	8	16	24	32	40	48	56	64	72	80
9	9	18	27	36	45	54	63	72	81	90
10	10	20	30	40	50	60	70	80	90	100

Fig. AS 1. 10 × 10 multiplication table.

+	1	2	3	4	5	6	7	8	9	10
1	2	3	4	5	6	7	8	9	10	11
2	3	4	5	6	7	8	9	10	11	12
3	4	5	6	7	8	9	10	11	12	13
4	5	6	7	8	9	10	11	12	13	14
5	6	7	8	9	10	11	12	13	14	15
6	7	8	9	10	11	12	13	14	15	16
7	8	9	10	11	12	13	14	15	16	17
8	9	10	11	12	13	14	15	16	17	18
9	10	11	12	13	14	15	16	17	18	19
10	11	12	13	14	15	16	17	18	19	20

Fig. AS 2. 10 × 10 addition table.

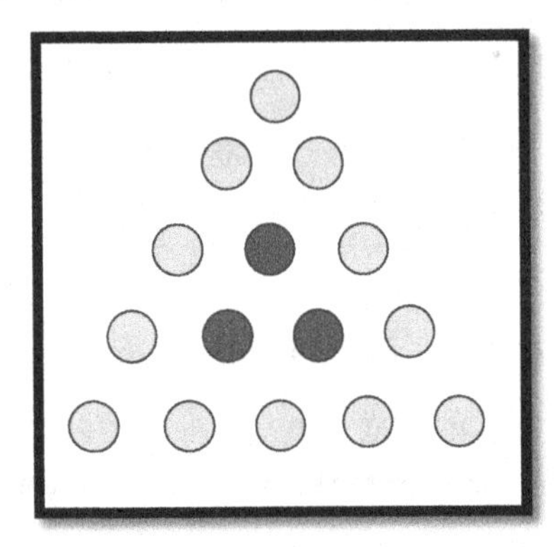

Fig. AS 3. Circles forming the fifth triangular number.

3) On the diagram of Fig. AS 3, do the following:

a. Count the total number of circles (within the five rows).

b. Count (from top to bottom) the number of circles in the first four rows.

c. Count (from top to bottom) the number of circles in the first three rows.

d. Find the difference between the numbers found in items b and c.

e. Count the number of circles shaded dark.

f. What is special about the numbers found in items a, b, c and e?

h. Show on the diagram of Fig. AS 3 that $15 = 3 \times (10 - 6) + 3$.

4) On the diagram of Fig. AS 4, do the following:

a. Count the total number of circles (within the six rows).

b. Count (from top to bottom) the number of circles in the first five rows.

c. Count (from top to bottom) the number of circles in the first four rows.

d. Find the difference between the numbers found in items b and c.

e. Count the number of dark shaded circles.

f. What is special about the numbers found in items a, b, c and e?

h. Show on the diagram of Fig. AS 4 that $21 = 3 \times (15 - 10) + 6$.

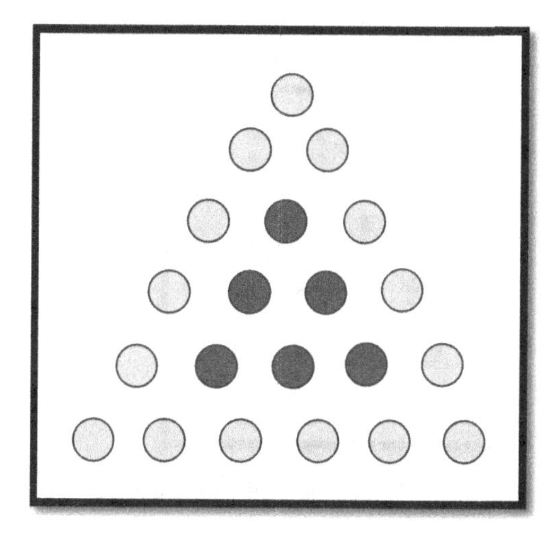

Fig. AS 4. Circles forming the sixth triangular number.

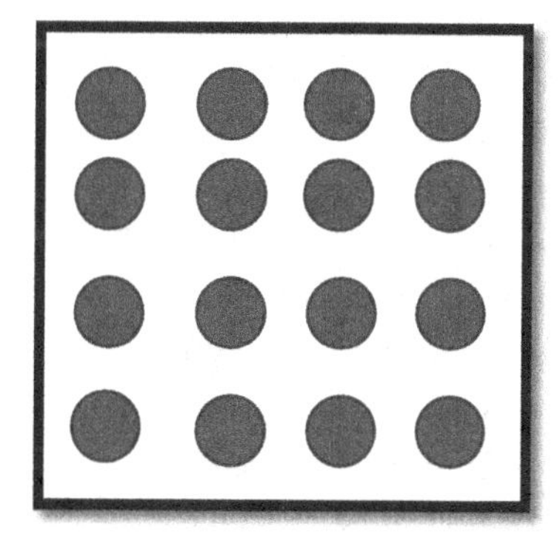

Fig. AS 5. Circles forming the fourth square number.

Fig. AS 6. Circles forming the fifth square number.

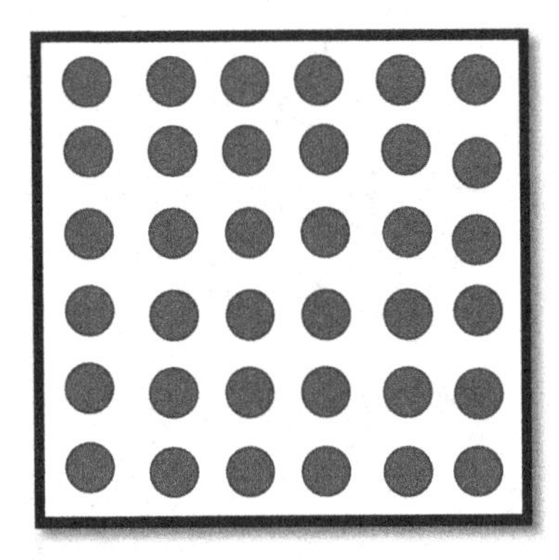

Fig. AS 7. Circles forming the sixth square number.

5) Let $a_n = 0.5n(n + 1)$, $n = 1, 2, 3, \ldots$.

Prove the identity $a_n = 3(a_{n-1} - a_{n-2}) + a_{n-3}$, $n \geq 4$.

6) Let $a_n = 0.5n(n + 1)$, $n = 1, 2, 3, \ldots$.

Prove the identity $a_n = 5(a_{n-2} - a_{n-3}) + a_{n-5}$, $n \geq 6$.

7) On the 4×4 square (Fig. AS 5) show that $16 = 1 + 3 + 5 + 7$.

8) On the 5×5 square (Fig. AS 6) show that $25 = 1 + 3 + 5 + 7 + 9$.

9) Using the square of Fig. AS 5, show that $16 = 3 \times (9 - 4) + 1$.

10) Using the square of Fig. AS 6, show that $25 = 3 \times (16 - 9) + 4$.

11) Using the square of Fig. AS 7, show that $36 = 5 \times (16 - 9) + 1$.

12) Prove the identity $3[(n - 1)^2 - (n - 2)^2] + (n - 3)^2 = n^2$, $n \geq 1$.

13) Prove the identity $5[(n - 2)^2 - (n - 3)^2] + (n - 5)^2 = n^2$, $n \geq 1$

14) The number of dots in the diagram of Fig. AS 8 develops through the sequence 1, 5, 12, 22, 35 – the first five pentagonal numbers (Chapter 5, Section 5.2, Remark 5.4). Using Fig. AS 8, show that the equalities $22 = 3 \times (12 - 5) + 1$ and $35 = 3 \times (22 - 12) + 5$ hold true. Compare the two equalities with those mentioned above in 3h and 4h.

15) Enter the numbers 1, 5, 12, 22, 35 into the input box of *Wolfram Alpha* to find the general term of pentagonal numbers.

16) Let $a_n = 0.5n(3n - 1)$, $n = 1, 2, 3, \ldots$. Find a_5.
Prove that $a_n = 3(a_{n-1} - a_{n-2}) + a_{n-3}$, $n \geq 4$.

17) Let $a_n = 0.5n(3n - 1)$, $n = 1, 2, 3, \ldots$. Find a_6. Show that $a_6 = 5(a_4 - a_3) + a_1$.
Prove that $a_n = 5(a_{n-2} - a_{n-3}) + a_{n-5}$, $n \geq 6$.

18) Enter the (hexagonal) numbers 1, 6, 15, 28, 45 into the input box of *Wolfram Alpha* to find the general term of this sequence.

19) Let $a_n = n(2n - 1)$, $n = 1, 2, 3, \ldots$. Show that $a_5 = 3(a_4 - a_3) + a_2$.
Prove that $a_n = 3(a_{n-1} - a_{n-2}) + a_{n-3}$, $n \geq 4$.

20) Let $a_n = n(2n - 1)$, $n = 1, 2, 3, \ldots$. Show that $a_6 = 5(a_4 - a_3) + a_1$.
Prove that $a_n = 5(a_{n-2} - a_{n-3}) + a_{n-5}$, $n \geq 6$.

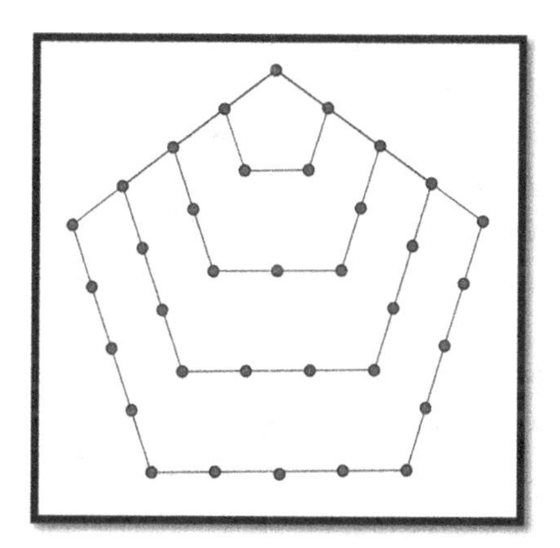

Fig. AS 8. Pentagonal numbers 1, 5, 12, 22, 35.

ACTIVITY SET 6

Unit fractions as benchmark fractions

1) Compare fractions (which one is bigger) using two-dimensional model (e.g., Chapter 7, Fig. 7.6): 1/2 or 1/3? Draw a picture (grid) demonstrating comparison.
2) Compare fractions (which one is bigger) using two-dimensional model (e.g., Chapter 7, Fig. 7.6): 2/5 or 3/7. Draw a picture (grid) demonstrating comparison.
3) Compare fractions (which one is bigger) using two-dimensional model (e.g., Chapter 7, Fig. 7.6): 3/4 or 4/5? Draw a picture (grid) demonstrating comparison.
4) How can one compare 3/4 and 4/5 using fraction circles without recourse to the two-dimensional model? Draw pictures (fraction circles) in making this demonstration. What is special about the fractions 3/4 and 4/5? Give three examples of similar pairs of fractions.
5) Divide 5 into 3 using divided-divisor context (Chapter 7) and pose a problem that leads to this division.
6) Divide 5 into 3 using part-whole context (Chapter 7) and pose a problem that leads to this division.
7) A unit fraction is a benchmark fraction. Recall that two unit fractions are considered consecutive benchmark fractions if their denominators differ by one. Place the fraction 3/16 between two consecutive benchmark (unit) fractions (Chapter 7, Section 7.8). Show your work. Use the context of dividing pizzas fairly when placing the fraction between two consecutive benchmark fractions.
8) Place the fraction 6/19 between two consecutive benchmark (unit) fractions (Chapter 7, Section 7.8). Show your work. Use the context of dividing pizzas fairly (Chapter 7, Section 7.9) when placing the fraction between two consecutive benchmark fractions.
9) Place the fraction 5/12 between two consecutive benchmark (unit) fractions (Chapter 7, Section 7.8). Show your work. Use the context of dividing pizzas fairly (Chapter 7, Section 7.9) when placing the fraction between two consecutive benchmark fractions.
10) Place the fraction 3/8 between two consecutive benchmark (unit) fractions (Chapter 7, Section 7.8). Show your work. Use the context of dividing pizzas fairly (Chapter 7, Section 7.9) when placing the fraction between two consecutive benchmark fractions.
11) Place the fraction 31/61 between two consecutive benchmark (unit) fractions (Chapter 7, Section 7.8). Show your work. Use the context of dividing pizzas fairly (Chapter 7, Section 7.9) when placing the fraction between two consecutive benchmark fractions.

12) Place the fraction 5/21 between two consecutive benchmark (unit) fractions (Chapter 7, Section 7.8). Show your work. Use the context of dividing pizzas fairly (Chapter 7, Section 7.9) when placing the fraction between two consecutive benchmark fractions.
13) Place the fraction 7/24 between two consecutive benchmark (unit) fractions (Chapter 7, Section 7.8). Show your work. Use the context of dividing pizzas fairly (Chapter 7, Section 7.9) when placing the fraction between two consecutive benchmark fractions.
14) Place the fraction 7/41 between two consecutive benchmark (unit) fractions (Chapter 7, Section 7.8). Show your work. Use the context of dividing pizzas fairly (Chapter 7, Section 7.9) when placing the fraction between two consecutive benchmark fractions.
15) Convert 3/16 into an Egyptian fraction (Chapter 7, Section 7.9). You may use *Wolfram Alpha* to check your work if you used the Greedy algorithm. Interpret each conversion as dividing pizzas among people fairly and find the number of pieces each person gets and the total number of pieces obtained through this division.
16) Convert 31/61 into an Egyptian fraction (Chapter 7, Section 7.9). You may use *Wolfram Alpha* to check your work if you used the Greedy algorithm. Interpret each conversion as dividing pizzas among people fairly and find the number of pieces each person gets and the total number of pieces obtained through this division.
17) Convert 5/21 into an Egyptian fraction (Chapter 7, Section 7.9). You may use *Wolfram Alpha* to check your work if you used the Greedy algorithm. Interpret each conversion as dividing pizzas among people fairly and find the number of pieces each person gets and the total number of pieces obtained through this division.
18) Convert 7/24 into an Egyptian fraction (Chapter 7, Section 7.9). You may use *Wolfram Alpha* to check your work if you used the Greedy algorithm. Interpret each conversion as dividing pizzas among people fairly and find the number of pieces each person gets and the total number of pieces obtained through this division.
19) Convert 7/36 into an Egyptian fraction (Chapter 7, Section 7.9). You may use *Wolfram Alpha* to check your work if you used the Greedy algorithm. Interpret each conversion as dividing pizzas among people fairly and find the number of pieces each person gets and the total number of pieces obtained through this division.
20) Convert 9/26 into an Egyptian fraction (Chapter 7, Section 7.9). You may use *Wolfram Alpha* to check your work if you used the Greedy algorithm. Interpret each conversion as dividing pizzas among people fairly and find the number of pieces each person gets and the total number of pieces obtained through this division.

ACTIVITY SET 7

From pre-operational to operational level of dealing with fractions

1) Complete operations step-by-step using pictures (diagrams) and describe what you see using fractional notation:

a. 3/8 + 1/4 (using fraction circles);

b. 3/5 + 4/15 (using fraction circles);

c. 5/6 + 1/2 (using fraction circles; is there more than one visually supported numeric answer?);

d. 5/6 – 1/2 (using fraction circles; is there more than one visually supported numeric answer?);

e. 5/6 – 1/12 (using fraction circles; is there more than one visually supported numeric answer?);

f. 5/6 – 1/3 (using fraction circles; is there more than one visually supported numeric answer?);

g. 2/3 + 1/6 (using rectangles);

h. 3/4 + 1/8 (using rectangles);

i. $1\frac{2}{3}-\frac{1}{6}$ (using set model; is there more than one visually supported numeric answer?);

j. $1\frac{5}{6}-\frac{1}{3}$ (using set model; is there more than one visually supported numeric answer?);

k. $1\frac{3}{4}-\frac{2}{8}$ (using set model; is there more than one visually supported numeric answer?);

2) Formulate a word problem that leads to an operation in the cases 1a – 1k.

3) If the strip ☐☐☐☐☐ is 5/3, draw the strip which is the whole.

4) If the strip ☐☐☐☐☐ is 5/4, draw the strip which is the whole.

5) If the strip ☐☐☐☐ is the whole, draw the strip which is 7/4 of the whole.

6) If the strip ☐☐☐☐ is 4/7, draw the strip which is the whole.

7) If the strip ☐☐☐☐☐ is 5/2, draw the strip which is the whole.

8) If the strip ☐☐☐☐ is the whole, draw the strip which is 5/4 of the whole.

9) If the strip ☐☐☐☐☐ is the whole, draw the strip which is 6/5 of the whole.

10) If the counters shown in Fig. AS 9 represent 2/5 of a set, how many counters are in the whole set?

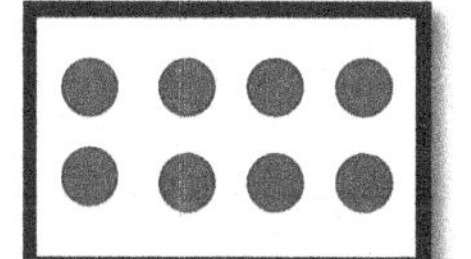

Fig. AS 9.

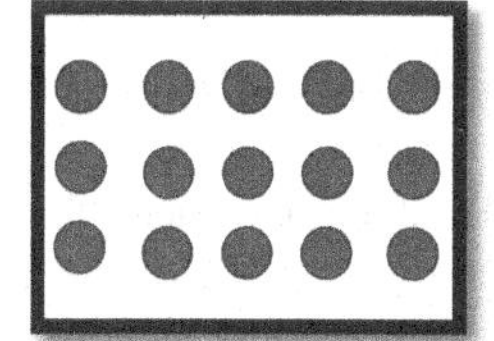

Fig. AS 10.

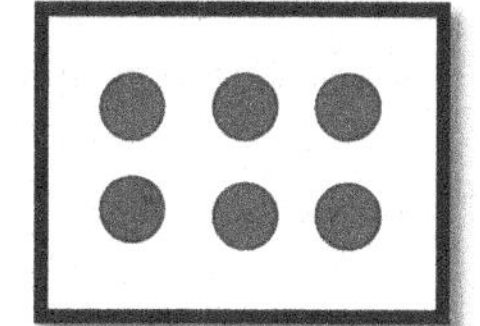

Fig. AS 11.

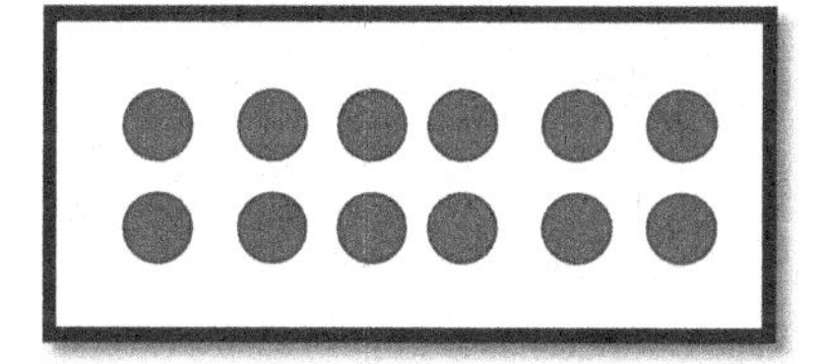

Fig. AS 12.

11) If the set of counters shown in Fig. AS 10 is one whole, how many counters are in 2/3 of the set?
12) If the set of counters shown in Fig. AS 11 is 2/3 of a set, how many counters are in the whole set?
13) If the counters shown in Fig. AS 11 represent 3/2 of a set, how many counters are in the whole set?
14) If the counters shown in Fig. AS 12 represent 2/5 of one whole, how many counters are in 1/2, 1/3, 1/5, 1/6, 7/10 and 9/15 of the whole?
15) Using a picture, show the meaning of reduction of the fraction 6/8 to the simplest form.

16) Using a picture, show the meaning of reduction of the fraction 4/12 to the simplest form.
17) Using a picture, show the meaning of reduction of the fraction 5/10 to the simplest form.
18) Using area model for fraction, complete multiplication using pictures (grids):
   a. $3\times\frac{2}{3}$ and show how formal multiplication stems from this picture.
   b. $3\times\frac{2}{5}$ and show how formal multiplication stems from this picture.
   c. $\frac{3}{5}\times 2$ and show how formal multiplication stems from this picture.
   d. $\frac{2}{5}\times 2$ and show how formal multiplication stems from this picture.
   e. $\frac{3}{8}\times\frac{2}{5}$ and show how formal multiplication stems from this picture.
   f. $\frac{5}{6}\times\frac{4}{5}$ and show how formal multiplication stems from this picture.
   g. $\frac{2}{3}\times\frac{5}{6}$ and show how formal multiplication stems from this picture.
   h. $\frac{4}{3}\times\frac{8}{5}$ and show how formal multiplication stems from this picture.
   i. $\frac{5}{3}\times\frac{4}{9}$ and show how formal multiplication stems from this picture.
   j. $\frac{7}{3}\times\frac{4}{5}$ and show how formal multiplication stems from this picture.
   k. Formulate a word problem that leads to an operation in each case, a – j.
19) Divide 9 by 3 through the *Invert and Multiply* rule and describe the division in terms of the change of unit (see Chapter 7, Section 7.6.1).
20) Divide 7 into 3 through the *Invert and Multiply* rule and describe the division in terms of the change of unit (see Chapter 7, Section 7.6.1).
21) Divide 6 into 5 through the *Invert and Multiply* rule and describe the division in terms of the change of unit (see Chapter 7, Section 7.6.1).
22) Divide 3/4 into 5/8 through the *Invert and Multiply* rule and describe the division in terms of the change of unit (see Chapter 7, Section 7.6.2).
23) Divide 5/6 into 3/4 through the *Invert and Multiply* rule and describe the division in terms of the change of unit (see Chapter 7, Section 7.6.2).
24) It is known that 4/7 of distance from A to B is equal to 11/5 miles. Find the distance from A to B using a picture and describe the process through the change of unit.

25) It is known that 6/5 of distance from A to B is equal to 7/8 miles. Find the distance from A to B using a picture and describe the process through the change of unit.

ACTIVITY SET 8

Pizza as a context for learning fractions

1) Divide three identical circular pizzas among four people fairly using dividend-divisor context for fractions, Egyptian fraction Greedy algorithm, and semi-fair division (see Chapter 7, Section 7.9). How many pieces of pizza do we have in each case? Draw pictures to show your solution.
2) Divide three identical circular pizzas among five people fairly using dividend-divisor context for fractions, Egyptian fraction Greedy algorithm, and semi-fair division (see Chapter 7, Section 7.9). Compare the number of pizza pieces obtained in each case. Draw pictures to show your solution.
3) Divide five identical circular pizzas among six people fairly using dividend-divisor context for fractions, Egyptian fraction Greedy algorithm, and semi-fair division (see Chapter 7, Section 7.9). How many pieces of pizza do we have in each case? Draw pictures to show your solution.
4) Divide five identical circular pizzas among eight people fairly using dividend-divisor context for fractions, Egyptian fraction Greedy algorithm, and semi-fair division (see Chapter 7, Section 7.9). How many pieces of pizza do we have in each case? Draw pictures to show your solution.
5) Divide five identical circular pizzas among nine people fairly using dividend-divisor context for fractions, Egyptian fraction Greedy algorithm, and semi-fair division (see Chapter 7, Section 7.9). How many pieces of pizza do we have in each case? Draw pictures to show your solution.
6) A pizza is cut into three different pairs of equal pieces: (1/4, 1/4), (1/6, 1/6) and (1/12, 1/12).
   a. Using computer program *Wolfram Alpha*, show that these pieces can be used separately to make the full pizza by repeating the first pair twice, the second pair three times, and the third pair six times.
   b. Find other ways to similarly cut pizza into three pairs of equal pieces out of which the whole pizza can be made.
   c. How can one cut pizza into four such pairs?

ACTIVITY SET 9

Word problems with fractions, percentages, and ratios

1) Jane eats 2/5 of a cup of cottage cheese per day. How much cup of cottage cheese does Jane eat over three days? Show your solution using a picture. Then provide a formal (step-by-step) arithmetical solution.
2) Ron bought four identical cakes and cut 2/3 from each cake to put on the table. How much cake did Ron put on the table? Show your solution using a picture. Then provide a formal (step-by-step) arithmetical solution.
3) Anna has six identical bottles of apple juice. If the serving glass is 3/4 of a bottle, how many servings can Anna make out of the bottles? Show your solution using a picture. Then provide a formal (step-by-step) arithmetical solution.
4) Three identical buckets are filled with water to the 2/3 capacity. How much water are in these buckets? Show your solution using a picture. Then provide a formal (step-by-step) arithmetical solution.
5) From each of the two identical cakes 2/5 was cut off. How much cake was cut off? Show your solution using a picture. Then provide a formal (step-by-step) arithmetical solution.
6) Jody has 5 pounds of flour. It takes 5/4 pounds to make a cake. How many cakes can Jody make? Solve the problem by using a picture and write a mathematical sentence that corresponds to your picture. Then provide a formal (step-by-step) arithmetical solution.
7) John has five identical bottles of orange juice. If the serving glass is 3/4 of a bottle, how many servings can John make out of the bottles? Show your solution using a picture. Then provide a formal (step-by-step) arithmetical solution.
8) Mary has $9\frac{3}{4}$ pounds of flour. It takes $3\frac{1}{4}$ pounds to make a cake. How many cakes can Mary make? Show your solution using a picture. Then provide a formal (step-by-step) arithmetical solution.
9) A turtle has crept the distance of $3\frac{1}{3}$ km. If it crept $\frac{2}{3}$ km per day, how many days did it take the turtle to creep the whole distance? Solve this problem as measurement and as a missing factor equation by using area model. Show your solution using a picture. Then provide a formal (step-by-step) arithmetical solution.
10) If Jonny's chore for a weekend is to paint 2/5 of their backyard fence and, on Saturday, Jonny did 3/8 of the chore, how much fence did Jonny paint on Saturday? Show your solution using a picture. Then provide a formal (step-by-step) arithmetical solution.

11) If a glass of milk is 4/5 full and a kitten was given 5/6 of this milk to drink, how much glass was given to the kitten? Show your solution using a picture. Then provide a formal (step-by-step) arithmetical solution.

12) A hospital has to pave two parking lots, A and B, shaped as squares of sides 80 meters and 120 meters, respectively. It is known that $62\frac{1}{2}\%$ of lot A and $58\frac{1}{3}\%$ of lot B need new pavement. Which parking lot needs more pavement and by how many square meters? Solve this problem using square grids. *Wolfram Alpha* can be used for calculations.

13) A college has to pave two parking lots, A and B, shaped as squares of sides $85\frac{2}{5}$ meters and $98\frac{2}{5}$ meters, respectively. It is known that 4/5 of lot A and 5/8 of lot B need new pavement. Which parking lot needs more pavement? *Wolfram Alpha* can be used for calculations.

14) Leila paid for a new laptop $1,500. It was on sale for 6.25% discount. How much money did Leila save through this discount? Show your solution using a picture. Then provide a formal (step-by-step) arithmetical solution.

15) Jeremy is one of only 6 boys in the class of 16 children. Find the ratio of boys to girls in this class? After Christmas break, Jeremy moved with his family to another state. Find the ratio of girls to boys in this class after the break?

16) The distance from A to B is 600 miles. It takes 12 hours to cover this distance by a car and 15 hours – by a truck. If the car and the truck started moving toward each other at the same time from A and B, respectively, in how many hours would they meet?

17) Solve problem 16) using a 60-cell grid (see Chapter 9, Section 9.5).

18) Solve problem 16) using a 40-cell grid (see Chapter 9, Section 9.5).

19) Solve problem 16) using a 30-cell grid (see Chapter 9, Section 9.5).

ACTIVITY SET 10

Informal geometry

1) How can one decide a numeric value of a fraction circle?
2) Give an example of a fraction circle that can be constructed out of two other fraction circles in more than two ways.
3) Do you know any connection that exists between the world of integer-sided rectangles and the world of fraction circles?
4) You have a closet filled with square-shaped desks each of which can seat four people. Preparing for a party, you arrange those desks (not necessarily all of them) to form a large rectangular table. Is it possible to have a table with as many seats as the number of desks used to create it?
5) You have a closet filled with square-shaped desks each of which can seat four people. Preparing for a party, you arrange those desks (not necessarily all of them) to form a large square-shaped table to seat a certain number of people. Is it possible to remove a few desks from this large table to still seat the same number of people? If so, how many desks may be removed?
6) Imagine a rectangle drawn on a square grid so that the vertices of the rectangle coincide with the points where the gridlines intersect. How can one find area of the rectangle without using a well-known formula for area?
7) Given a perimeter of an integer-sided rectangle, does there always exist more than one (integer-sided) rectangle with this perimeter? Why or why not? In the case of several rectangles, which rectangle has the largest area?
8) Given an area of an integer-sided rectangle, does there always exist more than one (integer-sided) rectangle with this area? Why or why not? In the case of several rectangles, which rectangle has the smallest perimeter?
9) Give an example when two quantities increase in such a way that percentage of one quantity within another quantity becomes smaller and smaller?
10) How can a triangle be connected to a straight angle?
11) You have square tiles the quantity of which is a prime number. How many different rectangles can you make out of these tiles (using all the tiles for a rectangle)?
12) You have square tiles the quantity of which is a square number. Can you make more than one rectangle out of these tiles (using all the tiles for a rectangle)? Why or why not?

13) You have square tiles the quantity of which is the square of a prime number. How many different rectangles can you make out of these tiles (using all the tiles for a rectangle)?

14) On a geoboard (or grid paper) construct (draw) a polygon with slanted sides (like the one shown in Fig. 10.10, Chapter 10) of area 6 square units (on the grid paper the vertices of the polygon coincide with the points where the gridlines intersect). Modify the polygon to make area 5 square units in as many ways as you can.

15) On a geoboard (or grid paper) construct (draw) a polygon with slanted sides (like the one shown in Fig. 10.10, Chapter 10) of area 6 square units (on the grid paper the vertices of the polygon coincide with the points where the gridlines intersect). Modify the polygon to make area 7 square units in as many ways as you can.

16) Is it possible to construct (draw) a polygon on a geoboard (or grid paper) with area numerically equal to perimeter? Why or why not?

17) On a geoboard (or grid paper) construct (draw) a polygon with area being numerically equal to half of perimeter. Can this be done in more than one way? Why or why not?

ACTIVITY SET 11
Combinatorics

1) A cafeteria offers four types of drinks – water, coffee, tea, juice – and three types of donuts – chocolate, strawberry, and blueberry. How many ways can one buy a pair of drink and donut? Answer this question by drawing a tree diagram.
2) A cafeteria offers four types of drinks – water, coffee, tea, juice – and three types of donuts – chocolate, strawberry, and blueberry. Yet only chocolate donut can be offered when one buys tea and only tea; that is, no other drink allows for a chocolate donut. How many ways can one buy a pair of drink and donut? Answer this question by drawing a tree diagram.
3) A cafeteria offers three types of drinks – water, coffee, tea – and four types of donuts – chocolate, glazed, strawberry, and blueberry. Only chocolate donut can be offered when one buys coffee and only coffee; that is, no other drink allows for a chocolate donut. Also, only glazed donut can be offered when one buys tea and only tea; that is, no other drink allows for a glazed donut. How many ways can one buy a pair of drink and donut? Answer this question by drawing a tree diagram.
4) At a book sale, Laura, without looking, reaches into a box filled with books priced $1 and $5. If Laura selects two books, how much money should Laura be prepared to pay?
5) At a book sale, John, without looking, reaches into a box filled with books priced $1, $2, and $5. If John selects two books, how much money should John be prepared to pay?
6) At a book sale, Ada, without looking, reaches into a box filled with books priced $3 and $5. If Ada selects three books, how much money should Ada be prepared to pay?
7) How many ways can one permute letters in the words KLEENEX, MIRRORBALL, ERRATA, CONTRIBUTION, RECURRENCE?
8) If Ron can enroll in a teacher education program in eight state and three private schools, how many choices of the program can Ron make?
9) Three different size towers are constructed out of six linking cubes. How many different orders can those towers be arranged?
10) Four different size towers are constructed out of ten linking cubes. How many different orders can those towers be arranged?
11) Six different size towers are constructed out of fifteen linking cubes. How many different orders can those towers be arranged?
12) How many ways can Jeannette put four rings on the fingers of the right hand excluding the thumb?

13) How many ways can Jeannette put five rings on the fingers of the right hand excluding the thumb and the index finger?
14) How many ways can one buy three drinks out of five types?
15) How many ways can one buy five drinks out of three types?
16) How many ways can three books be bought by four people?
17) How many ways can four books be bought by three people?
18) How many ways can seven books be bought by four people?
19) How many ways can the number 12 be decomposed in five addends arranged in all possible orders?
20) How many ways can the number 12 be decomposed in six addends arranged in all possible orders?
21) How many ways can the number 11 be decomposed in five addends arranged in all possible orders?
22) How many ways can the number 11 be decomposed in six addends arranged in all possible orders?
23) How many ways can the number 10 be decomposed in seven differently ordered addends?
24) How many ways can the number 10 be decomposed in nine addends arranged in all possible orders?
25) Seven identical cookies are put on three plates in all different orders, no plate having two cookies and no plate is empty. How many ways can this arrangement of cookies on plates be done?
26) Eight identical cookies are put on four plates in all different orders, no plate having three cookies and no plate is empty. How many ways can this arrangement of cookies on plates be done?
27) Ten identical cookies are put on five plates in all different orders, each plate having more than one cookie. How many ways can this arrangement of cookies on plates be done?

ACTIVITY SET 12

Probability and data management

1) Two dice are rolled. Find the theoretical probability that the number of spots on two faces is ten. Using *Wolfram Alpha*, show the sample space of this experiment in the form of an addition table. Construct a spreadsheet using the RANDBETWEEN function (see Chapter 13, Section 13.12) to find an experimental probability of this outcome when the dice are rolled 2000 times. Compare theory with the spreadsheet-based experiment.
2) Two dice are rolled. Find the theoretical probability that the number of spots on two faces is eight. Using *Wolfram Alpha*, show the sample space of this experiment in the form of an addition table. Construct a spreadsheet using the RANDBETWEEN function (see Chapter 13, Section 13.12) to find an experimental probability of this outcome when the dice are rolled 2000 times. Compare theory with the spreadsheet-based experiment.
3) Two dice are rolled. Find the theoretical probability that the number of spots on two faces is six. Using *Wolfram Alpha*, generate "6 addition table" to find the sample space of this experiment. Construct a spreadsheet using the RANDBETWEEN function (see Chapter 13, Section 13.12) to find an experimental probability of this outcome when the dice are rolled 2000 times. Compare theory with the spreadsheet-based experiment.
4) Three dice are rolled. Find the total number of outcomes of this physical experiment by using the Rule of Product (see Chapter 11, Section 13.1). What are the chances to have 10 spots on three faces if it is known that neither die had two spots when rolled? Construct a spreadsheet using the RANDBETWEEN function (see Chapter 13, Section 13.12) to find an experimental probability of this outcome when the dice are rolled 2000 times. *Suggestion.* Entering in the input box of *Wolfram Alpha* the command "solve over the integers a+b+c=10, 6>=a>=b>=c>0" yields three triples without the number 2. Permuting the elements in those triples (see Chapter 11, Section 11.3) yields the sample space of this physical experiment. Compare theory with *computational* experiment.
5) Three dice are rolled. Find the total number of outcomes of this physical experiment by using the Rule of Product (see Chapter 11, Section 13.1). What are the chances to have 11 spots on three faces if it is known that neither die had one spot when rolled? Construct a spreadsheet using the RANDBETWEEN function (see Chapter 13, Section 13.12) to find an experimental probability of this outcome when the dice are rolled 2000 times. *Suggestion.* Entering in the input box of *Wolfram Alpha* the command "solve over the integers a+b+c=11, 6>=a>=b>=c>1" yields four triples. Permuting the elements in those

triples (see Chapter 11, Section 11.3) yields the sample space of this physical experiment. Compare theory with the spreadsheet-based experiment.

6) A coin changing machine randomly changes a dollar into half dollars, quarters, and dimes. Represent a sample space of this experiment in the form of a table. Assuming that it is equally likely to get any combination of the coins, find the probability of having no dimes in the change.

7) A machine changed a half dollar coin into quarters, dimes, and nickels. Assuming that it is equally likely to get any combination of the coins, find the probability that
   (i) there are no nickels in the change,
   (ii) there are no dimes in the change,
   (iii) there are no quarters in the change.

8) A machine changed a dollar coin into half dollars, quarters, and dimes. Assuming that it is equally likely to get any combination of the coins, find the probability that
   (i) there are no dimes in the change,
   (ii) there are no quarters in the change,
   (iii) there are no half dollars in the change.

9) A half dollar coin is randomly changed into quarters, dimes, and nickels. Draw a sample space for this experiment in the form of a table.

10) A dollar coin is randomly changed into half dollars, quarters, and dimes. Draw a sample space in the form of a table.

11) Using ten square tiles, construct five towers and arrange them from the lowest to the highest. There may be more than one way to do that. Construct all such combinations of five towers using the ten tiles. Record your combinations (the sets of towers). For each set of towers find the mean number of stories, the median number of stories, and the mode number of stories. Describe what you have found. Assuming that it is equally likely to construct any set of five towers out of the ten tiles, what is the probability that towers are all the same size?

12) Using eight square tiles, construct four towers and arrange them from the lowest to the highest. There may be more than one way to do that. Record your combinations (the sets of towers). For each set of towers find the mean number of stories, the median number of stories, and the mode number of stories. Describe what you have found. Assuming that it is equally likely to construct any set of four towers out of the eight tiles, what is the probability that towers are all the same size?

13) Given the data set {1, 3, 6, 10, 10, 13, 15, 15, 15, 16}, determine the mean, median, mode, and range. Does the mode represent the data well?

14) Given the data set {1, 4, 9, 16, 25, 25, 32}, determine the mean, median, mode, and range. Which measure, mean or median, does represent the data better?

ACTIVITY SET 13

Open-ended questions

1) Why are questions considered as major means of learning mathematics?
2) Why do teachers need to know mathematics they are responsible for teaching beyond the grade level they are assigned to teach?
3) Why is mathematical knowledge of teachers especially important for the learning outcomes of students struggling with mathematics?
4) The Common Core State Standards [2010, p. 4, italics in the original] argued that teachers should be able "to justify, in a way appropriate to the student's mathematical maturity, *why* a particular mathematical statement is true or where a mathematical rule comes from". Do you agree with this position? Why or why not?
5) Why do we say it is equally likely to have either head or tail when tossing a coin? How do we know that? Indeed, one can toss a coin and have three tails in a row and still say that head and tail are equally likely.
6) What is the difference between informal geometry and formal geometry? Can you give an example of using informal geometry when solving a problem?
7) Can you give an example when solving a word problem using a picture yields an incorrect answer? What is the reason behind an erroneous result in that case?
8) What are the limitations of pictures as means of solving word problems?
9) Why do many teacher candidates prefer using decimal representation of numbers rather than their fractional representation?
10) What is your understanding of the effectiveness of mathematics in solving real-life problems?
11) How do you understand the meaning of the words "appropriate use of technology" in the context of teaching and learning mathematics?
12) What is the difference between procedural problem solving and conceptual problem solving?
13) What is the relation between fair sharing and division as an arithmetical operation?
14) What are the strategies used in the teaching of mathematics to demonstrate that certain statement is not true?
15) What is the meaning of the word counterexample? Give a real-life application of using a counterexample as a reasoning tool.
16) What are the roles of counterexamples in the teaching of mathematics?
17) There are three sets of counters out of which an image of a certain figure has to be created. Three cases are possible: two images can be created, no image can be created, one

image can be created. In which order would you present the three cases to young children? Explain your rationale.

18) How would you teach mathematics to make it likable?
19) Does teaching problem solving through multiple strategies make mathematics more likable? Why or why not?
20) Why are fractions difficult to teach and learn? Why are they more difficult than integers?
21) Why is the concept of numeric equivalence applicable to fractions and not applicable to integers?
22) What is a geometric interpretation of the common denominator of two fractions? Linguistically, how can you explain the meaning of the two words, numerator and denominator, in relation to fractions?
23) Why do we use a grid, rather than a number line, to compare two common fractions (that is, to determine the larger/smaller one)?
24) How can one turn an integer into a unit fraction? Try to think about a real-life situation when an integer turns into a unit fraction.
25) Why is it more difficult to add fractions than to multiply fractions? Indeed, whereas one has to find the common denominator of two (or more) fractions in order to add them, multiplication of fractions is pretty straightforward by multiplying numerators and denominators.
26) Have you ever noticed how squares and, consequently, square numbers – 1, 4, 9, 16, 25, 36, ... – develop? If you have not, try to notice now and share your observations.
27) How could you use manipulatives (e.g., square tiles) to explain the development of squares and square numbers?
28) Where do mathematical formulas come from? Do mathematicians just see them in their dreams? Or do they see something else?
29) What is your experience with the multiplication table?
30) How can *Wolfram Alpha* be used with elementary schoolchildren?
31) What is your experience with solving a mathematical problem in more than one way?
32) What is your opinion about encouraging visual thinking in the context of mathematical problem solving?
33) Can you give an example when visual thinking in mathematics may be misleading?
34) Do you have experience in solving procedural tasks conceptually? Why is such experience important?
35) What is the difference between productive and reproductive thinking?

36) How can a teacher explain to a third-grade student that the chances to roll an even number when rolling a fair (six-sided) die are equal to 1/2?

37) As a teacher, how would you stimulate the emergence of insight in your students?

38) What may be a real-life situation when you have to decompose an integer in two or more addends?

Bibliography

Abramovich, S. (2005). Early algebra with graphics software as a Type II application of technology, *Computers in the Schools*, 22(3/4), pp. 21–33.

Abramovich, S. (2007). Uncovering hidden mathematics of the multiplication table using spreadsheets, *Spreadsheets in Education*, 2(2), pp. 158–176.

Abramovich, S. (2010) *Topics in Mathematics for Elementary Teachers: A Technology-Enhanced Experiential Approach*, (Information Age Publishing, Charlotte, NC).

Abramovich, S. (2016). Exploring Mathematics with Integrated Spreadsheets in Teacher Education, (World Scientific, Singapore).

Abramovich, S. (2017). *Diversifying Mathematics Teaching: Advanced Educational Content and Methods for Prospective Elementary Teachers*, (World Scientific, Singapore).

Abramovich, S. (2020a). Paying attention to students' ideas in the digital era, *The Teaching of Mathematics*, 23(1), pp. 1–16.

Abramovich, S. (2020b). Pizzas, Egyptian fractions and practice-fostered mathematics, *Open Mathematical Education Notes*, 10(2), pp. 33–43.

Abramovich, S. (2021). Using Wolfram Alpha with elementary teacher candidates: From more than one correct answer to more than one correct solution, *MDPI Mathematics*, 9, 2112, 18 pages, https://doi.org/10.3390/math9172121, accessed on January 19, 2022.

Abramovich, S. and Brouwer, P. (2003). Revealing hidden mathematics curriculum to pre-teachers using technology: The case of partitions, *International Journal of Mathematical Education in Science and Technology*, 34(1), pp. 81–94.

Abramovich, S. and Brouwer, P. (2011). Where is the mistake? The matchstick problem revisited, *PRIMUS – Problems, Resources, and Issues in Mathematics Undergraduate Studies*, 21(1), pp. 11–25.

Abramovich, S., and Cho, E. K. (2008). On mathematical problem posing by elementary pre-teachers: The case of spreadsheets, *Spreadsheets in Education*, 3(1), pp. 1–19.

Abramovich, S., Stanton, M. and Baer, E. (2002). What Are Billy's chances? Computer spreadsheet as a learning tool for younger children and their teachers alike, *Journal of Computers in Mathematics and Science Teaching*, 21(2), pp. 127–145.

Aleksandrov, A. D. (1963). A general view of mathematics. In A. D. Aleksandrov, A. N. Kolmogorov and M. A. Lavrent'ev (Eds), *Mathematics: Its Content, Methods and Meaning*, (MIT Press, Cambridge, MA) pp. 1–64.

Apostol, T. M. (1967). *Calculus, Vol. 1: One-variable Calculus, with Introduction to Linear Algebra*, (Wiley, Hoboken, NJ).

Arnheim, R. (1969). *Visual Thinking*, (University of California Press, Berkeley and Los Angeles, CA).

Association of Mathematics Teacher Educators. (2017). *Standards for Preparing Teachers of Mathematics*, https://amte.net/standards, accessed on January 19, 2022.

Baker, J. and Sugden, S. J. (2003). Spreadsheets in education – The first 25 years, *Spreadsheets in Education*, 1(1), pp. 18–43.

Becker, J. P. and Selter C. (1996). Elementary school practices. In A. J. Bishop, K. Clements, C. Keitel, J. Kilpatrick and C. Laborde (Eds), *International Handbook of Mathematics Education,* (Kluwer, Dordrecht, The Netherlands) pp. 511–564.

Beiler, A. H. (1964). *Recreations in the Theory of Numbers: The Queen of Mathematics Entertains*, (Dover, New York, NY).

Blum, W. and Niss, M. (1991). Applied mathematical problem solving, modeling, applications, and links to other subjects – state, trends and issues in mathematics instruction, *Educational Studies in Mathematics*, 22(1), pp. 37–68.

Bouygues, H. L. (2019). *Does educational technology help students learn?* (The Reboot Foundation, Paris, France), https://reboot-foundation.org/wp-content/uploads/_docs/ED_TECH_ANALYSIS.pdf, accessed on January 19, 2022.

Canobi. K. H. (2005). Children's profiles of addition and subtraction understanding, *Journal of Experimental Child Psychology*, 92(3), pp. 220–246.

Catalan, E. (1884). Notes sur la théorie des fractions continues et sur certaines séries [Notes about the theory of continued fractions and certain series]. *Mémoires de L'Académie Royale des Sciences, des*

Lettres et des Beaux-Arts de Belgique, Tome XLV, (F. Hayez, Imprimeur de L'Académie Royale, Bruxelles, Belgium).

Chace, A. B., Manning, H. P. and Archibald, R. C. (1927). *The Rhind Mathematical Papyrus, British Museum 10057 and 10058, vol. 1*, (The Mathematical Association of America, Oberlin, OH).

Chang, K. Y. (2013). Teaching and learning practices in mathematics classrooms: focused on geometry. In J. Kim, I. Han, M. Park and J. K. Lee (Eds), *Mathematics Education in Korea*, (World Scientific Singapore) pp. 153–172.

Common Core State Standards. (2010). *Common Core Standards Initiative: Preparing America's Students for College and Career*, http://www.corestandards.org, accessed on January 19, 2022.

Conference Board of the Mathematical Sciences. (2001). *The Mathematical Education of Teachers*, (The Mathematical Association of America, Washington, DC).

Conference Board of the Mathematical Sciences. (2012). *The Mathematical Education of Teachers II*, (The Mathematical Association of America, Washington, DC).

Connell, M. L. (2001). Actions upon objects: A metaphor for technology enhanced mathematics instruction. In D. Tooke and N. Henderson (Eds), *Using Information Technology in Mathematics*, (Haworth Press, Binghamton, NY) pp. 143–171.

Dai, Q. and Cheung, K. L. (2015). The wisdom of traditional mathematical teaching in China. In L. Fan, N.-Y. Wong, J. Cai and S. Li. (Eds), *How Chinese Teach Mathematics: Perspectives from Insiders*, (World Scientific, Singapore) pp. 3–42.

Department for Education (2013, updated 2021). National Curriculum in England: Mathematics Programmes of Study, Crown copyright. https://www.gov.uk/government/publications/national-curriculum-in-england-mathematics-programmes-of-study, accessed on January 19, 2022.

Department of Basic Education. (2018). *Mathematics Teaching and Learning Framework for South Africa: Teaching Mathematics for Understanding*, (The Author, Private Bag, Pretoria).

Dewey, J. (1933). *How We Think*, (D. C. Heath and Company, New York, NY).

Dewey, J. (1938). *Experience and Education*, (MacMillan, New York, NY).

Dyson, G. (2012). *Turing's Cathedral: The Origins of the Digital Universe*, (Pantheon Books, New York, NY).

Felmer, P., Lewin, R., Martínez, S., Reyes, C., Varas, L., Chandía, E., Dartnell, P., López, A., Martínez, C., Mena, A., Ortíz, A., Schwarzc, G. and Zanocco, P. (2014). *Primary Mathematics Standards for Pre-Service Teachers in Chile*, (World Scientific, Singapore).

Freudenthal, H. (1978). *Weeding and Sowing*, (Kluwer, Dordrecht, The Netherlands).

Gattegno, C. (1971). *Geoboard Geometry*, (Educational Solutions Worldwide, New York, NY).

Gelman, R. and Meck, E. (1983). Preschooler's counting: principles before skills, *Cognition*, 13(3), pp. 343–359.

Hofstadter, D. R. (1985). *Metamagical Themas: Questing for the Essence of Mind and Pattern*, (Basic Books, New York, NY).

Hwang, H. and Han, H. (2013). Current national mathematics curriculum. In J. Kim, I. Han, M. Park and J. K. Lee (Eds), *Mathematics Education in Korea*, (World Scientific, Singapore) pp. 21–42.

Isaacs, N. (1930). Children's why questions, In S. Isaacs, *Intellectual Growth in Young Children*, (Routledge & Kegan Paul, London, England) pp. 291–349.

Katz, V. (Ed.). (2007). *The Mathematics of Egypt, Mesopotamia, China, India and Islam. A Sourcebook*, (Princeton University Press, Princeton, NJ).

Kaufman, E. L., Lord, M. W., Reese, T. W. and Volkmann, J. (1949). The discrimination of visual number, *The American Journal of Psychology*, 62(4), pp. 498–525.

Kaur, B. (2009). Performance of Singapore students in trends in international mathematics and science studies. In W. K. Yoong, L. P. Yee, B. Kaur, F. P. Yee and N. S. Fong (Eds), *Mathematics Education: The Singaporean Journey*, (World Scientific, Singapore) pp. 439–463.

Kline, M. (1985). *Mathematics for the Non-Mathematician*, (Dover, New York, NY).

Koshy, T. (2001). *Fibonacci and Lucas Numbers with Applications*, (Wiley, New York, NY).

Kuijt, I. (Ed.) (2002). *Life in Neolithic Farming Communities: Social Organization, Identity and Differentiation*, (Kluwer, New York, NY).

Kuo, E., Hull, M.M., Gupta, A. and Elby, A. (2013). How students blend conceptual and formal mathematical reasoning in solving physics problems, *Science Education*, 97(1), pp. 32–57.

Luchins, A. S. and Luchins, E. H. (1970). *Wertheimer's Seminars Revisited: Problem Solving and Thinking*, volume I, (SUNY at Albany Faculty-Student Association, Albany, NY).

Maddux, C. D. and Johnson, D. L. (2005). Type II applications of technology in education: New and better ways of teaching and learning, *Computers in the Schools*, 22(1/2), pp. 1–5.

Ministry of Education, Singapore. (2012). *Mathematics Syllabus, Primary One to Four*, (The Author, Curriculum Planning and Development Division),https://www.moe.gov.sg/docs/default-source/document/education/syllabuses/sciences/files/mathematics-syllabus-(primary-1-to-4).pdf, accessed on January 19, 2022.

Ministry of Education Singapore. (2020). *Mathematics Syllabuses, Secondary One to Four*, (The Author, Curriculum Planning and Development Division,), https://www.moe.gov.sg/-/media/files/secondary/syllabuses/maths/2020-express_na-maths_syllabuses.pdf?la=en&hash=95B771908EE3D777F87C5D6560EBE6DDAF31D7EF, accessed on January 19, 2022.

Nam, J. Y., Kwon, S., Yim, I., and Park, K. S. (2013). Teaching and learning practices in mathematics classrooms: focused on problem solving, In J. Kim, I. Han, M. Park and J. K. Lee (Eds), *Mathematics Education in Korea*, (World Scientific, Singapore) pp. 198–219.

National Council of Teachers of Mathematics. (2000). *Principles and Standards for School Mathematics*, (The Author, Reston, VA).

National Curriculum Board. (2008). *National Mathematics Curriculum: Framing Paper*, (The Author, Australia), http://www.acara.edu.au/verve/_resources/National_Mathematics_Curriculum_-_Framing_Paper.pdf, accessed on January 19, 2022.

Neuwirth, E. and Arganbright, D. (2004). *The Active Modeler: Mathematical Modeling with Microsoft Excel,* (Brooks/Cole, Toronto, Canada).

New York State Education Department. (1998). *Mathematics Resource Guide with Core Curriculum*, (The Author, Albany, NY).

Ontario Ministry of Education. (2020). *The Ontario Curriculum, Grades 1–8, Mathematics (2020)*, http://www.edu.gov.on.ca, accessed on January 19, 2022.

Rudman, P. S. (2007). *How Mathematics Happened. The first 50,000 years*, (Prometheus Books, Amherst, NY).

Piaget, J. and Inhelder, B. (1963). *The Child's Conception of Shape*, (Routledge & Kegan Paul, London, England).

Pólya, G. (1981). *Mathematical Discovery: On Understanding, Learning and Teaching Problem Solving*, Combined edition, (Wiley, New York, NY).

Power, D. J. (2000). *DSSResources.com*. http://dssresources.com/history/sshistory.html, accessed on January 19, 2022.

Serrazina, L. and Rodrigues, M. (2015). Additive adaptive thinking in 1st and 2nd grades pupils. In K. Krainer and N. Vondrová (Eds), *Proceedings of the Ninth Congress of the European Society for Research in Mathematics Education* (pp. 368–374), Prague, Czech Republic, htttp://hal.archives-ouvertes.fr/hal-01286846, accessed on January 19, 2022.

Smith, D. E. (1924). The first printed arithmetic (Trevisto, 1478), *Isis*, 6(3), pp. 311–331.

Takahashi, A., T. Watanabe, Yoshida, M., and McDougal, T. (2004). *Elementary School Teaching Guide for the Japanese Course of Study: Arithmetic (Grade 1–6)*, (Global Education Resources, Madison, NJ).

Van de Walle, J. A. (2001). *Elementary and Middle School Mathematics: Teaching Developmentally* (4th edition), (Addison Wesley Longman, New York, NY).

Van der Waerden, B. L. (1961). *Science Awakening*, (Oxford University Press, New York, NY).

Van Hiele, P. M. (1986). *Structure and Insight*, (Academic Press, Orlando, FL).

Vygotsky, L. S. (1978). *Mind in Society*, (Harvard University Press, Cambridge, MA).

Vygotsky, L. S. (2001). *Lectures on the dynamic of child development* [*Лекции по Педологии*], (In Russian), G. S. Korotaeva, T. I.

Zelenina, A. M. Gorfunkel, A. V. Zhukova, A. N, Utehina, N. V. Mahan'kova, T. I. Belova and L. I. Maratkanova (Eds), (Publishing House "Udmurt State University Press": Izhevsk, Russian Federation), https://www.marxists.org/russkij/vygotsky/pedologia/lektsii-po-pedologii.pdf, accessed on January 19, 2022.

Wertheimer, M. (1959). *Productive Thinking*, (Harper & Brothers, New York, NY).

Wu, Y, and Wong, K. Y. (2009). Understanding of statistical graphs among Singapore secondary students. In W. K. Yoong, L. P. Yee, B. Kaur, F. P. Yee and N. S. Fong (Eds), *Mathematics Education: The Singaporean Journey*, (World Scientific, Singapore) pp. 227–243.

Young-Loveridge, J. (2002). Early childhood numeracy: Building an understanding of part-whole relationships, *Australian Journal of Early Childhood*, 27(4), pp. 36–42.

Index

A

abstraction, 1, 5, 95, 103, 109, 116, 117
activity-based learning, 169
addition table, 55, 60, 62, 65, 67, 206, 230, 231, 245–247
addition, 15, 16, 20, 34, 166
additive partition, 11
algebraic thinking, 55
algorithm, 31, 105, 131, 148
ambiguous case, 36
analysis, 161, 162
area, xiv, 38, 55, 102, 108, 158, 165, 168, 177, 179, 180, 184, 210, 260, 265
Aristotle, 7
arithmetic sequence, 40, 62
array, 22, 27, 180, 181, 183
Association of Mathematics Teacher Educators, 3, 19, 21, 39, 60, 97, 105, 150, 170, 176, 212
ATLT rule, 78–80, 83, 86, 88, 248, 249
Australia, 179
average, 214
axiom, 162

B

base-ten number, 2
benchmark fractions, 121, 122, 257
big idea, 41, 97, 118, 143, 184,
binary classification, 11
Bolyai, 162

C

calculator, 39, 241
cardinality, 2, 5, 15
Catalan, 89
chances, 199, 207, 209
change of unit, 112, 113, 260, 261
Chile, 15, 39, 159, 170, 207
classification, 10
collateral learning, 120
color, 11
combination with repetition, 194
combination without repetition, 194
Common Core State Standards, viii, ix, 11, 31, 33, 42, 55, 100, 103, 105, 118, 121,133, 143, 166, 182, 193, 198, 199, 272,
common sense, 31, 119
commutative property, 17, 25, 29, 35, 45, 58, 77, 237
comparison, 9, 17
completion, 17
computational fluency, 169
computing, 37
conceptual coherence, 118
conceptual knowledge, 42
conceptual shortcut, 37, 38, 40–43, 80, 88, 91, 59, 237–239
conceptual understanding, 23, 99, 137, 139, 148, 150, 153, 169,
conceptualization, 88
concrete situation, 46
Conference Board of the Mathematical Sciences, 7, 60, 69, 93, 100, 106, 123, 131, 150, 176, 179, 180, 202, 39,
Confucius, 3
conservation of a number, 5
contextual coherence, 70, 71
contextual meaning, 43
contextualization, 1, 17, 43, 104, 122
convention, 36
counting skills, 4
counting, 2, 4–6, 9, 47
17, 162, 165, 180, 185
Cuisenaire rods, 165

D
decimal fraction, 131
decomposition, 15, 17, 50, 53, 80, 123, 163, 182, 194,
decontextualization, 33, 93, 101, 104, 122
deduction, 158
denomination, 101, 117
denominator, 102
Dewey, 59, 60, 120, 177
diagonal, 159
didactical coherence, 70
difference, 145
digit, 2, 24
discovery, 82
distributive property, 106
dividend, 132
dividend-divisor context, xiii, 95–97, 119, 125, 128, 131, 145, 256, 262
division, 28, 29, 34, 97
divisor, 132
dodecagon, 173
doubles, 15
doubling phenomenon, 23–25

E
Egyptian civilization, 2, 167
Egyptian fraction, 124, 125, 127–129, 257, 262
Egyptian papyrus roll, 2, 167
enclosure, 160
England, 2, 3, 15, 105
equally likely outcomes, 199, 204, 209
equation, 42, 43, 44, 72, 80, 88, 92, 115, 116, 147, 153
equilateral triangle, 157, 219
equivalence, xii, 97, 99, 120
equivalent fractions, 97, 99
even and odd numbers, 6
even numbers, 8, 15, 65, 67, 86, 245–246
experiment, 200, 201, 204, 210, 269
experimental probability, 200, 269
explanation, 13

F
factor, 22, 30
factorial, 191
fair coin, 199, 204
Fibonacci numbers, 89
Fibonacci, 1, 124
Fibonacci-like sequence, 89
finger multiplication, 24
finger rings, 46, 49, 197, 198
fluency, 169
formal deduction, 161
fraction circle, 217, 218, 265
fractions, 11, 93, 209

G
Gauss, 40
generalization, 12, 22, 58, 88, 89, 181
geoboard, xiv, 159, 165, 166, 168, 169, 179, 266
Geometer's Sketchpad, viii, ix, xiv, 174, 175, 217, 224
geometric interpretation, 44
geometric series, 123
geometric thinking, xiv, 160, 162, 163
geometrization, 22
Gestalt psychology, 166
gnomon, 7, 28, 41, 58, 60, 62
Golden Ratio, xiii, 143, 150
graph, 44, 212
Greedy algorithm, 124, 125, 257, 262
guessing, 6

H
hands-on, x, xi, xiv, 25, 29, 30, 80

Heron of Alexandria, 7
hexagon, 11, 173
histogram, xv, 212

I
image, 7, 40, 42, 84, 101, 106, 217, 272
improper fraction, 102, 107
inequalities, 121
infinite descent, 81
infinity, 5, 30
informal deduction, xiv, 161, 174, 175, 180, 181, 184
informal geometry, 157, 166, 167, 272
information, 13
insight, 37, 58, 74
intuition, 157
invariance, xiii, 5, 97, 143, 174, 175, 184
inverse proportional relationship, 150
Invert and Multiply rule, 111–113, 260
isosceles trapezoid, 11
isosceles triangle, 2, 157
iterating, 118, 120

J
Japan, 4, 22, 93, 150, 166, 167

L
Lamé, 89
least common multiple, 98
Leibniz, 11
likelihood, 200, 202
line plot, xv, 212
linear equation, 44
linear unit, 159, 165, 180
Lobachevsky, 162
long division, xiii, 31, 33, 131, 132
Lucas, 89

M
mean, xv, 214, 215, 270, 271
measurement model, 30, 109, 112, 114, 138
measurement, 2, 3, 4, 95, 118, 144
measures of central tendency, 215
measuring, 9
median, xv, 214, 215, 270, 271
metacognition, 7, 70
metacognitive development, 51
Ministry of Education Singapore, 5, 7, 11, 28, 51, 59, 60, 70, 81, 88, 91, 93, 104, 161, 169, 179, 198
misconception, xiv, 137, 139, 140, 150
missing factor, 28, 109, 110, 113, 115
mixed fraction, 98, 102
mode, xv, 214, 215, 270, 271
multiplication table, 22, 24, 25, 28, 31, 55, 57, 58, 229, 245, 246, 247
multiplication, 20, 21, 23, 29, 34, 186, 237, 260

N
National Council of Teachers of Mathematics, 189, 199
Neolithic culture, 3
non-collinear points, 157
non-terminating decimal, 133
number line, 86, 118, 119, 120
numerator, 102
numeric misconception, 9
numerical coherence, 70, 71

O
odd numbers, 7, 8, 28, 65, 67, 86, 245, 246, 247
OEIS, 76, 83, 90
one-to-one correspondence, 9, 10, 118

Ontario Ministry of Education, 7, 8, 42, 60, 71, 81, 179, 186,
operation, 15, 16, 28, 30, 35, 95–97, 104
order of operations, 34
outcome, 199, 204, 269

P
parity, 123
partition model, 30, 99, 109, 112, 115
partition, 95
part-whole context, 95, 96, 143, 256
Pascal, 61
Pascal's triangle, 61, 205
pattern blocks, 11
pattern, 12, 13, 14, 27, 52, 61
pedagogical coherence, 70
pentagon, 64, 143, 150, 221
pentagonal numbers, 62, 255
percent, xiii, 134
perimeter, xiv, 55, 158, 177, 179, 180, 184, 266
permutation, 191–193
pet store mathematics, 91
Piaget, 160
Pick's formula, xiv, 168, 169
place value chart, 18, 240, 241
place value, 2, 3, 18
plane, 157, 172
Plato, 6
Plutarch, 179
polygon, 168, 169, 266
positional notation, 2
prediction, 202
prime number, 265, 266
prism, 180
probability, 199, 200, 201, 210, 270
problem posing, 17, 20, 37, 60, 69, 70, 120, 167, 272
procedural, 4
product, 24, 55
productive thinking, 137, 67
proof, 185
proper fraction, 102
proportion, 147, 148, 150, 154–156
proportional relationship, 150
Pythagoras, 3
Pythagorean theorem, 159
Pythagorean triples, xiv, 159

Q
quadrilateral, 158, 161
quotient, 32, 132, 145

R
randomness, 200, 201
range, 200, 243
rate, xiv, 152
ratio, xiii, 100, 123, 143, 147, 159, 199, 200, 211, 264
rational number, 134
rectangle, 38, 158, 159, 265
rectangles with a hole, 177
rectangular, 153
referent unit, 93
regrouping, 19, 37
remainder, 132
repeated addition, 100, 186, 237
repetition, 12
rhombus, 11, 161
rigor, 162
Rule of Product, 186, 188, 189, 191, 210, 269
Rule of Sum of Products, 187, 191
Rule of Sum, 187, 188, 189

S
sample space, xv, 202, 203, 205, 210, 270
scalene triangle, 157
semi-fair division, 125, 126, 128, 129, 262
semi-perimeter, 55, 182
set model, 97, 258

set, 5
shape, 11
size, 11
South Africa, 9, 39, 150, 161, 166, 177
South Korea, 69, 172
spreadsheet, viii, ix, 66, 72, 74, 79, 82, 87, 177, 201, 209, 211, 229, 231, 248, 249, 269
square number, 41
square tiles, 177
square, 161, 173, 220
starts, xiv
stem and leaf plot, xv, 213
straight angle, 157, 223, 265
straight line, 44
subitizing, 6
subtraction, 9, 17, 18, 20, 34, 166
summation, 7
symbolic description, 12
symbolic, 104
symmetry, 23, 161

T
take away, 17
tally marks, 2
technology, ix, 37, 69, 70, 74, 81, 120
terminating decimal, 131, 133
tessellation, 11, 171–173, 175, 222, 224
theorem, 162
theoretical probability, 269
tile, 47, 112
topological primacy, 160
tower, 13, 14, 47, 48, 50, 51, 53, 112, 137
train, 53
tree diagram, xiv, xx, 185–189, 191, 267
trial and error, 88, 238
triangle inequality, ix, 157
triangle, 11, 157
triangular numbers, 7
two-sided counters, 11

U
unbiased die, 199, 204
uniform movement, 114, 154
unit cubes, 180
unit fractions, 93, 94, 96, 121–123, 256
unit rate, 153
unit square, 165, 166, 180
unit, xiii, 3, 4, 96, 101, 109, 110, 112, 137, 139, 147

V
van Hiele model, 162, 163, 180–182, 184
visual patterns, 12
visual, 9, 39, 55, 84, 104, 139, 162, 163, 209
visualization, 160, 162, 209
volume, 180
Vygotsky, 100, 164

W
W^{4}S principle, 45, 46, 101
well-ordering principle, 81
Wertheimer, 166
whole number, 5
whole, 93, 94, 97, 104, 106
209, 258, 259
whole-part, 16
Wolfram Alpha, viii, ix, 56, 61, 75, 124, 129, 133, 142, 153, 159, 197, 245, 246, 257, 262, 264, 269

Printed in the United States
by Baker & Taylor Publisher Services